The Staircase of Life

The Story of the Origin and Evolution of Life and Humankind

GERARD ALEXANDER
WILLIGHAGEN

Gerard Alexander Willighagen

"The urge to comprehend the story of life and humankind leads to a deeper understanding of our existence's rarity, improbability, and significance."

Gerard Alexander Willighagen has a Master of Business Administration (MBA) in International Management from the University of Applied Sciences TIO. Despite the absence correlation between his literary pursuits and his academic background, Willighagen possesses a notable level of expertise and interest in the fields of anthropology and astrobiology. His initial literary work, *The Great Filter*, serves as the overarching focus of his present and forthcoming endeavours.

In Advance…

Thank You
For Reading

"I'd love to hear from you!"

—Gerard Alexander Willighagen

To request permissions, contact the publisher at nobis.publishing.house@gmail.com

ISBN paperback: 9789083335407

First paperback edition 2023

Written and edited by Gerard Alexander Willighagen

CONTENTS

PART I   THE BIOLOGICAL REVOLUTION

PART II   THE COGNITIVE REVOLUTION

ACKNOWLEDGEMENTS

These last few years, I have grown so captivated by the thought that our minds are fundamentally those of the Stone Age and that every living creature is connected to the first sparks of life that my interest in the evolution and history of life and humankind has increased dramatically. This book was created under the intense rigours of a busy schedule and different work as an MBA student. Although I am solely responsible for the flaws in this book, I have relied extensively on a number of individuals and sources without whom this book could not have been written.

First and foremost, I would want to give Yuval Noah Harari the most outstanding credit for being the source of the majority of my inspiration and knowledge. In his famous and insightful book *Sapiens: A Brief History of Humankind*, Yuval Noah Harari examines the development of *Homo sapiens* and its impact on human history. I have used, interpreted, and articulated with caution the book's structure, topics and knowledge in my own words.

Secondly, I owe Bill Bryson great credit. In his book *A Short History of Nearly Everything*, Bryson explores the history of everything, from the Big Bang to the origin of life. I owe him a lot for using the insightful knowledge and structure of his last chapters on the origin and evolution of life in my own interpretation. In addition, I would like to give big credit and acknowledgement to Robert Winston for the fact that I have attentively utilised a fascinating section of his book *Human Instinct*, which discusses the Stone Age instinctive characteristics of humankind. Lastly, I would like to thank

Stefan Milosavljevich, who produces the finest archaeological documentaries on the Internet, from which I have utilised and articulated in my own words a great deal of inspiration, information, and topics about the stories and rise of *Homo sapiens*.

With these acknowledgements, it is safe to assume that the majority of the greatest ideas and theories in this book are not mine, but I accept full responsibility for any inaccuracies, interpretational failures, and assumptions. In addition to being included in the text (with a square number), the list of noteworthy sources and credits that I have used, interpreted, and articulated in my own words may be found at the back of the book.

Science is the process of seeking answers and determining the best approach to finding them. It cannot be flawless since it is performed by people in an imperfect environment. Science doesn't always discover the answers and is often inconclusive and indecisive—which is important to keep in mind while reading the theoretical pieces. Thus, there is a potential that the theories in this book will be proven erroneous in decades or perhaps even years, which is precisely what makes science so fascinating.

It's also worth noting that some of the most intriguing theories—such as the origin, evolution and history of life and *Homo sapiens*—may be considered controversial by some. While I don't expect this book to answer all of your questions about the origin and evolution of life and the rise and evolution of the human species, I do hope this book aids you in an honest, engaging and respectful manner to fully appreciate and comprehend who we truly are and how we got here.

INTRODUCTION

From the world's most crowded metropolises, where life never seems to slow down, to the most remote, smallest, and most peaceful villages. From thousands of exotic, unique languages and dialects to worldwide-spoken languages, such as Mandarin, English, Spanish and Arabic. From Christians, Muslims, Hindus, Buddhists and agnostic theists, to atheists and the still searching. And from the endless distinct cultures, traditions, rituals, and skin colours to different nations, morals and values. Everyone is extremely unique, yet so the same. In times of polarisation, globalisation, digitalisation, and warfare, *Homo sapiens* may overlook that everyone has something very special in common: our evolution, our history.

Ever since the birth of our history, *Homo sapiens* have yearned to comprehend the true meaning of existence and our purpose and place in this hostile Universe. It is estimated that there are trillions upon trillions of planets in the Universe capable of supporting life, suggesting that there should be much room for its development. However, when we look up at the night sky, we do not see anything. The Universe seems to be empty and dead. Yet, the only place that appears to be flourishing and has successfully climbed the evolutionary staircase of life is the eight-billion *Homo sapiens*-rich planet Earth, which (for now) suggests that the rest of the Universe is devoid of life.

However, as I stated in my previous book (*The Great Filter*), perhaps there is an evolutionary step on the staircase of life that may have occurred on our habitable, oxygen- and water-rich planet that is exceedingly improbable to utterly impossible elsewhere in the Universe. Perhaps our planet is the only place that contains life, like a tiny, fragile candle in a vast, dark, cold room. Or maybe this is untrue, and we just haven't seen it yet.

Nonetheless, it is crucial to thoroughly understand the evolutionary staircase of life before scrutinising the endless potential habitable places to solve whether we are alone or not. Thus, what fundamental principles influence the origin and evolution of life? What did life's story look like? How did *Homo sapiens* ascend to the top so that you could experience existence and read this book? And is the evolutionary story of life and *Homo sapiens* really that memorable?

I'll give you a sneak peek: our story is indeed exceptional, notable, and utterly intriguing. Nevertheless, before we embark on our story, the emergence and rise of *Homo sapiens*, the dawn of civilisation and its imperial history to the assembly of today's modern world, let's start at the bottom of the evolutionary staircase, where things are inconspicuous and small, yet teeming with action. Or should I say, microscopic small?

VII

PART I

THE BIOLOGICAL REVOLUTION

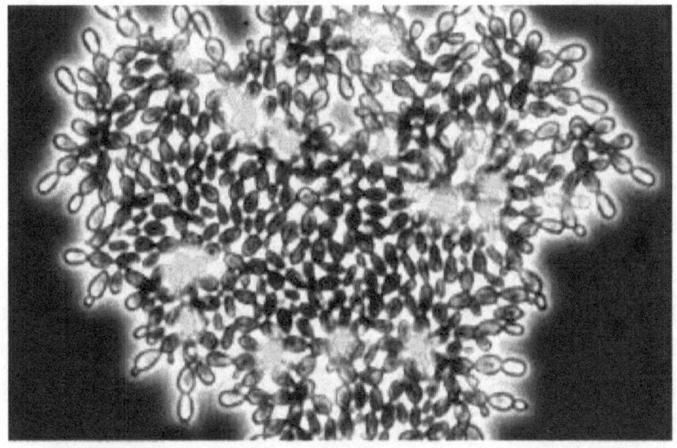

0) Scientists "evolved" Brewer's yeast, a single-celled organism, into multi-celled clusters that cooperate, reproduce, and adapt to their surroundings, thus recreating a crucial evolutionary phase in the development of life on Earth.

The Origin and Evolution of Life

1. THE ORIGIN OF LIFE

In 1953, Stanley Miller, a PhD student at the University of Chicago, filled two flasks: one with water to represent a primordial ocean and the other with a combination of methane, ammonia, and hydrogen sulphide gases to simulate the Earth's early atmosphere. Miller then connected them with rubber tubing and sparked a series of electrical sparks to simulate lightning. The fluid gathered at the trap became pink after a day, and after a week of continuous operation, the solution was dark red and thick. Miller discovered five amino acids in the fluid using paper chromatography.[1]

The Miller experiment was an intriguing chemical experiment that reproduced the circumstances assumed to exist on the early, primordial Earth at the time and explored the chemical genesis of life under those circumstances. After Miller died in 2007, experts examined sealed bottles kept from the original tests and discovered that well over 20 distinct amino acids were created in Miller's initial studies, many more than Miller originally reported.[2]

The Synthesis of Life

Despite a half-century of further research, we are no closer to synthesising life now than we were in 1953. Many scientists now believe that the early atmosphere was a far less reactive mixture of nitrogen and carbon dioxide, rather than Miller's gaseous soup. So far, replicating Miller's inspired efforts with these more difficult inputs has given only one fairly rudimentary amino acid. As a result, the synthesis of amino acids is not the main issue. In any case, the main issue is protein synthesis.

When you chain amino acids together, you get proteins (and you need a lot of them to have a functioning protein). Nobody knows for sure, but there might be several million different types of proteins in the human body, and each one is a tiny wonder. Interestingly, proteins should not exist according to all probability rules. To construct a protein, you must arrange amino acids (also known as the building blocks of life) in a certain order, just like you would arrange letters to spell a word. The order (also known as the Proteomic Code) controls how certain protein-protein interactions interact with distinct amino acids.

The issue is that the amino acid alphabet words are sometimes quite lengthy. To spell collagen, the name of a kind of protein, you must arrange eight letters correctly. To produce collagen, however, 1,055 amino acids must be arranged in the correct order. But here's the obvious but vital point: you're not going to make it. It appears spontaneously and without guidance, which is where the improbabilities arise. According to Brill Bryson, in his book *A Short History of Nearly Everything*,[0] the chances of a 1,055-sequence molecule like collagen spontaneously self-assembling are negligible. "No matter how much you'll try, it's simply not going to happen."

Yet we're talking about many thousands, if not millions, different types of proteins, each one unique and, as far as we know, necessary for the existence of life. And it gets much more improbable from there. A functioning protein must not only assemble amino acids in the correct order, but

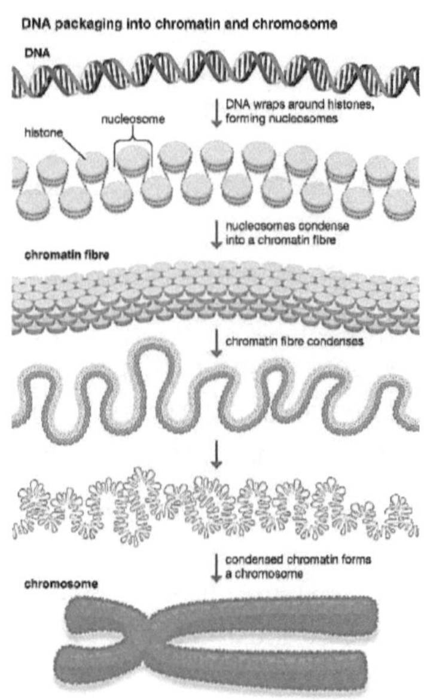

1) DNA is wrapped around proteins known as histones to form nucleosomes. These units condense into a chromatin fibre, which further condenses into a chromosome.

it must also engage in chemical origami and fold itself into a very specific form. Even if it achieves this level of structural complexity, it is no longer beneficial for life since it cannot replicate itself: DNA is required for reproduction.[0]

There's even more to discuss. DNA, proteins, and other components of life could not exist without some type of membrane to keep them contained. No atom or molecule has ever successfully attained life on its own. These many elements can only produce life when they join together in a cell. They are little more than "floating" chemicals outside the cell. However, the cell is useless without the chemicals.

It's no surprise that we call it the "miracle of life." The incredible and interesting thing is that we are only just now beginning to comprehend the miracle. So, what accounts for all of this marvellous complexity? One option is that it is not as beautiful as it appears at first look. What if, instead of bursting into existence, proteins mysteriously evolved?

Chemical reactions of the type associated with life are really rather prevalent. It may be beyond our abilities to produce them in a lab, as Stanley Miller demonstrated, but the Universe, strangely enough, does so easily. Many molecules in nature link together to create lengthy chains known as polymers ("polymer" is derived from the Greek words "polus," meaning "many", and "meros", meaning "part").

Interestingly, sugars constantly come together to form starch. Crystals can do a variety of lifelike things, such as replicate, respond to environmental stimuli, and take on pattern complexity.[3] Of course, they have never attained life itself, but they have repeatedly demonstrated that complexity is actually a natural, accidental, and completely everyday phenomenon. There may or may not be life in the Universe, but there is no lack of structured self-assembly, from the fascinating symmetry of snowflakes to Saturn's magnificent rings.[0]

This natural urge to self-assemble is so strong that many scientists now argue that the evolution and development of life is more inevitable than we realise. According to Christian de Duve, a Belgian biologist and Nobel winner, "it is a compulsory manifestation of matter, obliged to occur when the conditions are appropriate." De Duve estimated that such circumstances would occur a million times in each galaxy.[4]

Yet every scenario you've ever heard or read about the circumstances required for life involves liquid water. Earth had essentially little atmosphere and a molten surface when it was formed from a heated mix of gases and minerals some 4.6 billion years ago.[7] However, when the Earth cooled, an atmosphere formed, primarily from gases emitted by volcanoes. It included hydrogen sulphide, methane, and 10 to 200 times the amount of carbon dioxide found in today's atmosphere. After around half a billion years, the Earth's surface cooled and consolidated sufficiently for water to remain liquid.

These "liquidy" terms may be found everywhere, from Charles Robert Darwin's "warm little pond" to the bubbling marine vents that are nowadays the most common theories for the origin of life. Hydrothermal vents have recently sparked a great deal of curiosity among those bubbling sea vents because they provide a favourable environment for the creation of the earliest protocells (a self-organised, endogenously ordered, spherical collection of lipids proposed as a stepping-stone toward the origin of life).

Deep underwater hydrothermal vents spring up (mainly at the rifts dividing tectonic plates where ocean water seeps into the Earth's mantle and releases vents at high temperatures and pressure). It was always a puzzle how amphiphiles (a chemical compound possessing both water-loving and fat-loving properties) were formed, a critical component for protocells. However, the fluid flowing out of the vent is divided by the surrounding seawater by crystallising a thin layer of iron-nickel sulfide-rich minerals, forming the vent chimney. The vent fluid is hot, alkaline, and

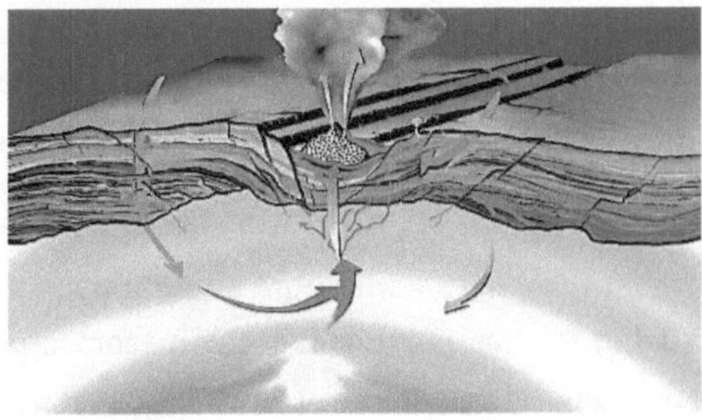

2) Cold seawater seeps into cracks in the seafloor and can be heated up to a ferocious 400° C by interacting with magma-heated subsurface rocks. The heat stimulates chemical reactions that extract minerals and chemicals from the rocks, after which the mineral- and chemical-rich soup rises to the surface through vent holes.

has high levels of dissolved molecular hydrogen. On the other hand, the surrounding seawater is cold, acidic, and has high levels of dissolved carbon dioxide. This reduction of carbon dioxide forms formic acid (systematically named methanoic), the starting point for creating lipid bilayers. The formic acid decomposes under hot hydrothermal vent conditions, forming carbon monoxide. After this, the components will react to the single-chain amphiphiles, providing fatty acids and long-chain alcohols in various chain lengths. The amphiphiles will line up to give their most stable configuration and form lipid bilayer structures. If the amphiphiles are collectively large enough, they will pinch off, forming a daughter cell, which is not a living system but represents an essential component of all life as we know it.

One of the greatest surprises in the Earth sciences of recent decades has been the discovery of how early in Earth's history life arose. Before the 1950s, it was believed that life was about 600 million years old. Yet, by the 1970s, scientists had a feeling it might go back 2.5 billion years. But the current date of 3.85 billion years is staggeringly early (the Earth's surface only became solid about 3.9 billion years ago!).

Because life arose so quickly, some authorities believe it must have had help (some argue maybe a lot). The theory and belief that Earthly life came from space has a remarkably lengthy and, at times, illustrious history. According to recent research published in *Nature*, meteorites may have delivered the foundation of life's genetic code. Analyses of three meteorites reveal that nucleobases, which are essential components of DNA, might have originated in space and subsequently dropped to Earth, providing the raw material for the genesis of life.

The complexities of how these molecules form in interstellar space are still widely debated, because, according to Charlotte Bays, PhD student of Planetary Science at Royal Holloway, University of London, and the Natural History Museum, London, "there are many different ways they could be synthesised." Surprisingly, *Nature* research finds nucleobases in these meteorites, which make up the backbone of RNA and DNA. While guanine and adenine were previously discovered, the discovery of cytosine, uracil, and thymine completes the set required for base pair formation. The presence of these complex biological molecules in meteorites eliminates some of the difficulties related to their formation on Earth, since the delivery of these first building blocks would have opened the way for crucial bio-relevant molecules such as DNA and RNA to develop later.[6]

The most remarkable truth in biology, arguably the most striking and amazing fact we know, is that whatever triggered life to begin, it only happened once. Everything that has ever existed, from plants and animals to fungi and bacteria, can be traced back to this primal twitch. Some little speck of molecules fidgeted to life in the impossibly distant past. It absorbed some nutrients, slowly pulsed, and lived for a short time. This might have happened before, possibly several times. But this ancestral packet went over and above by cleaving itself and producing an heir. A little bundle of genetic material was passed from one living entity to the next and hasn't stopped moving ever since. For every one of us, including you, it was the moment of creation: "the Big Birth."[0]

The Big Birth

This heir was probably as simple as life gets, but it was still life. It survived. It grew and spread. And that eventually led to you. But what the Earth was like back then is extremely debatable. However, there had to be something that suited life (otherwise, you wouldn't be here). However, it would not have been a very pleasant habitat for humans. If you stepped into ancient Archaic Earth from a time machine, you'd quickly retreat because there was no more oxygen to breathe on Earth back then than there is on Mars now.

According to researchers at the NAI's New York Center for Astrobiology at Rensselaer Polytechnic Institute, Earth was filled with poisonous vapours from hydrochloric and sulfuric acids that were strong enough to eat through clothes and burn human skin. The chemical stew in the atmosphere would have prevented much sunlight from reaching the Earth's surface. What little light there was would be illuminated temporarily by bright and occasional lightning.[7] So, in short, it was indeed the Earth, but an Earth we wouldn't recognise as our own.

For two billion years, the only forms of life in this Archaen world were unicellular bacterial organisms. They thrived, reproduced, and swarmed despite the fact that the environment was hellish from a human standpoint. Nonetheless, it achieved all of this without demonstrating any intent to progress to a higher, more sophisticated level of life. Cyanobacteria, or blue-green algae, learnt to tap into a freely accessible resource during their first billion years of life: the hydrogen that occurs in abundance in water. These cyanobacteria absorbed water molecules, consumed hydrogen, expelled oxygen, and developed photosynthesis. According to Lynn Margulis, an American evolutionary scientist, photosynthesis is undoubtedly the most significant single metabolic breakthrough in the history of life on Earth. Surprisingly, it was developed by bacteria rather than plants.

By studying ancient rocks, researchers have determined that sometime between 2.5 and 2.3 billion years ago, Earth underwent what scientists call the "Great Oxidation Event." The Great Oxidation Event (GOE), also known as the Oxygen Catastrophe or the Oxygen Revolution, was a historical period when the amount of oxygen in the Earth's atmosphere and shallow oceans first increased.[8] As cyanobacteria flourished, the Earth began to fill with oxygen, much to the chagrin of species that found it toxic (which in those days was virtually all of them).

By using the abundant potential of oxygen in respiration, these cyanobacteria were able to thrive in the deadly oxygen environment.[9] Many existing anaerobic organisms on Earth were wiped out by the abrupt infusion of poisonous oxygen into an anaerobic environment. According to Malcolm S.W. Hodgskiss of Stanford University's

Department of Geological Sciences, although the Great Oxidation Event is not considered a major biotic event, one of his studies suggests that the decline in gross primary productivity during this transition is the largest extinction event in the planet's history. [10]

But what was the secret to the success of these oxygen-using organisms? Simply because of two benefits: First, oxygen was a more efficient way of producing energy; Second, it eliminated rival species. Cyanobacteria were a smashing success. Initially, the extra oxygen they created did not concentrate in the atmosphere, but instead mixed with iron to form ferric oxides, which sunk to the bottom of primaeval oceans. The globe actually rusted for millions of years (a phenomenon vividly recorded in the banded iron deposits that supply so much of the world's iron ore today). Nothing more than this happened for tens of millions of years, and there would be little evidence of life if you returned to that early Proterozoic Earth. Perhaps in hidden puddles, you'd come across a bit of live goo or a layer of glossy greens and browns on shoreline rocks, but otherwise, life was apparently undetectable. [8]

However, something more striking became visible around 3.5 billion years ago. Visible structures began to form wherever the oceans were shallow. The cyanobacteria grew very sticky as they went through their chemical routines, and that stickiness trapped microparticles of dust and sand, which joined together to form very bizarre but substantial structures: stromatolites (layered sedimentary formations). Scientists had known about stromatolites from fossil formations for many years, but the discovery of a cluster of living stromatolites in Shark Bay on Australia's remote

3) Modern stromatolites in Shark Bay, Western Australia.

northwest coast in 1961 took them by surprise. This was so surprising that it took experts several years to realise what they had discovered.[11]

However, it took so long for life to become complex because the Earth had to wait for simpler organisms to properly oxygenate the atmosphere. It took nearly two billion years (almost 40% of Earth's lifetime) for oxygen levels in the atmosphere to reach more or less present levels. However, once the oxygen stage was set, an altogether new sort of cell developed. According to Kartik Aiyer, PhD microbiologist and researcher at Aarhus University in Denmark, "one with a nucleus and several little entities generally known as organelles" (from a Greek term meaning "little tools").[12]

The process is supposed to have begun when a reckless or brave prokaryotic unicellular bacterium (also known as a prokaryote, single-celled organism, which means "pre-nucleated") invaded or was caught by another prokaryotic bacterium, which turned out to suit them both (it didn't kill it). Since the beginning of unicellular existence, prokaryotic unicellular bacteria constantly devour and fight

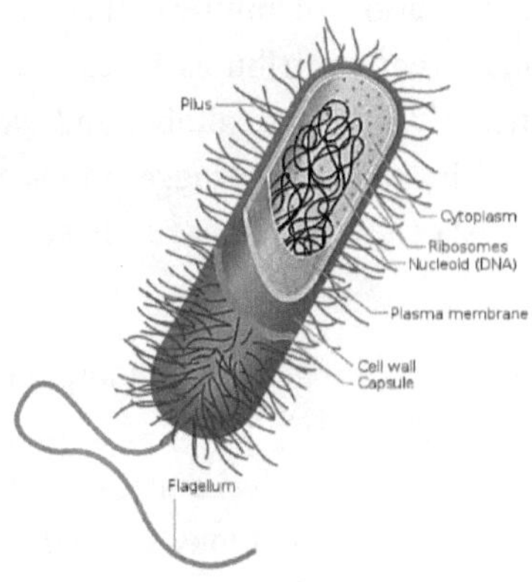

4) Able to thrive in extremely harsh environments, from the snowy surface of Antarctica to the hot undersea hydrothermal vents, most prokaryotic cells can divide every 24 hours, or even faster with an adequate supply of food.

each other for resources and energy. However, a union was made this time, and the imprisoned prokaryotic bacterium no longer had to worry about the harmful environment or other prokaryotic bacteria, and it used the "free," plentiful energy (provided by the other prokaryotic bacterium) to enhance itself. The imprisoned prokaryotic bacteria cell grew bigger as a result of this "amalgamation" and eventually developed into the cell's powerhouse: the mitochondrion. Ultimately, this mitochondrial invasion (or "endosymbiotic event," as scientists call it) enabled complex multicellular life to develop (a similar "amalgamation" generated chloroplasts, which allow plants to photosynthesize).

Thus, oxygen produced by cyanobacteria was responsible for changes in the composition of the Earth's atmosphere, the advent of aerobic metabolism, and, eventually, the evolution of multicellularity. Oxygen is the principal chemical that contributes to Earth's current state, which is significantly more hospitable and elegant than the early Earth. It is not an exaggeration to state that cyanobacteria are responsible for your existence and current form.

Mitochondria utilise and manipulate oxygen to release energy (adenosine triphosphate, called ATP), which is then used as a source of chemical energy throughout the cell. Without this clever approach, today's life on Earth would be little more than a muck of basic bacteria. Mitochondria are quite small. Yet, almost all of the nutrients you take in are used to nourish them. As a result, it is often referred to as a cell's "powerhouse" or "power plant." You won't be able to survive reading more than a few pages of this book without them.

Yet, despite this, mitochondria still have their own DNA, reproduce at a different period than their host cell, appear like bacteria, divide like bacteria, and occasionally respond to antibiotics in the same manner as bacteria do. They don't even share the same genetic language as the cell they dwell in. It's like having a stranger in your home, except this stranger has been there for a billion years.[13]

This new cell is known as a eukaryote (which means "fully nucleated" in Greek). Grypania, the earliest known eukaryotes, were found in iron deposits in Michigan in 1992. Even with a half-billion-year-old history, such fossils have only been discovered once. Eukaryotes were larger, eventually up to ten times larger than prokaryotes, and had up to 10,000 times more DNA.

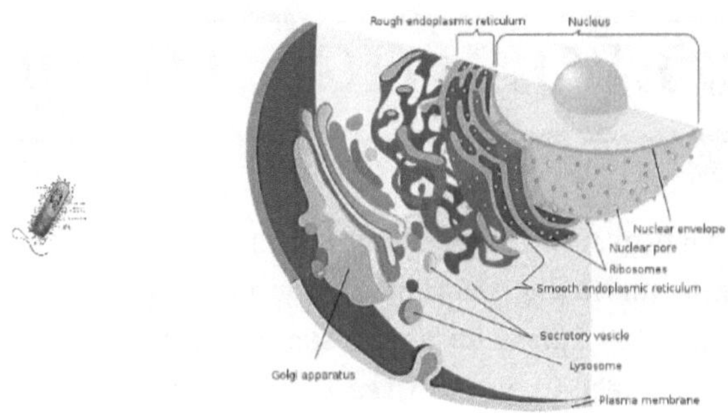

5) Eukaryotic cell (right) compared to a prokaryote (left).

As Bryson put it simply: "Life gradually developed into a system dominated by two sorts of organisms: those that exhale oxygen and those that take it in".[0] However, interestingly, at least 20 times in the history of life—and possibly many more—these single-celled eukaryotes have evolved into multicellular forms that are larger than those of their ancestors. Only in a few of these instances multicellularity "erupted," leading to the complex organisms known as plants, animals, fungi, and some forms of algae.

Nonetheless, with these pioneer phases of the evolutionary staircase discussed, an intriguing question arises: How could eukaryotes have generated virtually infinite biodiversity and species from only a small number of these "eruptions"? How could it have possibly led to us? And how could it have ultimately led to your existence?

As we'll see in the following chapters, despite the fact that scientists only recently decoded how this process actually works, it is far more interesting, fascinating, and sophisticated than any scientists could have ever imagined.

2. EVOLUTION THEORY

From a prokaryotic unicellular bacterium to you, every living being is markedly different. A group of fish, birds, bears, or humans may appear to be the same at first glance, but with closer inspection, we can see that they all have somewhat marginal distinct physical features and characteristics. Some are larger, stronger, shorter, or more adventurous than others. Others are more lethargic, aggressive, or intelligent. Others may have even fewer noticeable differences, such as a slightly darker or lighter skin tone, eye colour, body odour, or a greater metabolism.

Nature's Uniqueness

DNA is responsible for these minor differences. Every living creature is made up of cells that have a nucleus (containing chromosomes that hold the DNA). This is the information, instruction, and order code for cells. Everything about a creature is determined by it, from eye colour to character. And it is the DNA that distinguishes each creature.

One crucial aspect contributes to the vast diversity of DNA: an excess of offspring. In nature, organisms generally produce far more offspring than is required for their species' survival. Because many offspring die at an early age, there is an excess of offspring production. As a result, the greater the number of offspring produced, the greater the likelihood of the species' survival.[0] This overproduction of offspring contributes to the growing variety: The more offspring produced, the more minor changes in DNA are possible. Nature truly prefers these little deviations. One of life's essential rules is that nothing is perfect. Perfection is unattainable. Life would not exist without imperfections.

The second significant source of individual differences is heredity. Heredity transmits DNA to offspring, resulting in recombination and mutation. Recombination is the random mixing of DNA that occurs during mating. When animals mate, their genes are recombined twice. The first time occurs as a result of gametes such as sperm and egg cells, and the second occurs when the organisms mate. When a male inseminates a female, each parent contributes 50% of the offspring's DNA. In other words, each parent contributes 50% of their unique traits and characteristics to their offspring, resulting that the offspring's DNA is even more randomised. With each generation, the randomness grows.

Mutations are the third primary source of uniqueness. Mutations are unpredictably occurring alterations in DNA that can be regarded as copying mistakes. Toxins can trigger these mutations, such as chemicals or (ionising) radiation. Ionising radiation can strike the DNA molecule directly, like a bullet, ionising and damaging it. Assume that you would read the DNA segment—CATCATCATCATCAT—in groups of three: CAT-CAT-CAT-CAT-CAT. It can be shown that the segment's reading framework is altered if a nucleotide base, like T, is added after the first C: CTA-TCA-TCA-TCA-TCA. The coding sequence of amino acids is changed after the insertion. However, the original reading frame will be preserved in the remainder of the sequence if a total of three nucleotides are added or removed. Frameshift mutations occur when a nucleotide or

6) As a result of mutations (changes in the DNA sequence), the shell phenotypes (observable characteristics or traits of an organism through the expression of an organism's genetic code and the influence of environmental factors) within the bivalve mollusk species *Donax variabilis* display an array of colour and patterning.

set of nucleotides is added or removed in a way that disrupts the original sequence. It's fascinating to consider the fact that mutation rates in bacteria and other microbes are often lower than in more complicated animals like our species, *Homo sapiens*.[0]

These alterations are frequently detrimental to the species and can lead to ailments such as cancer. Other common mutation examples in humans are Angelman syndrome, colour blindness, Down syndrome, haemophilia, Turner syndrome, and many more.

Yet, interestingly, mutations can have a neutral or even favourable effect. The blue eye colour is a neutral result of mutation for humans, and our rich colour vision is a favourable mutation for humans. Humans have trichromatic vision, which means we can see three different colours: red, green, and blue. Unlike humans, many animals can only perceive certain colours due to having a dichromatic or monochromatic vision.

However, in the context of our story, mutation may have been the driving force behind the emergence and evolution of life. The term "background radiation" refers to a steady stream of ionising radiation from a wide range of environmental sources. There are four main types of natural radiation, as identified by the United Nations Scientific Committee on the Effects of Atomic Radiation (UNSCEAR): cosmic radiation, terrestrial radiation, inhalation of naturally existing radionuclides, and ingestion of naturally occurring radionuclides. These types of natural radiation bombard every living thing on Earth continually.

The mutations caused by cosmic radiation are among the most intriguing and likely causes of the emergence and evolution of complex life on Earth. There is, of course, continuous cosmic radiation that hits Earth; nevertheless, the most exciting and dangerous cosmic radiation comes not in a steady stream but rather as a massive, very unlikely storm: the explosion of a star, also known as a supernova (plural: supernovae).

These explosions from massive, dying stars are so luminous that they may temporarily outshine all the stars in their home galaxy for a short time. Many supernovae have been seen by mankind, but they have always been thousands of light-years away. Many of them were brighter than the Moon, and they lasted a few weeks before fading away. Yet, about once every few million years, a star within 300 light-years will appear in the sky and cast a twilight-like glow throughout the night. While this is too faint and far away to do any damage to humanity, it nevertheless has the potential to impact our planet.

Supernovae have undoubtedly showered Earth with radioactive debris, as shown by the 1999 discovery of considerable amounts of a weakly radioactive variant of iron known as iron-60 in deep-ocean rocks. Compared to the massive quantities of iron-60 produced by supernovae, the quantity produced by other natural processes is just a few per cent to a tenth as high. Supernovae that erupt within 30–45 light-years of Earth have been theorised to have devastating repercussions for life on Earth due to the high-energy particles they release. Previous research, however, suggested that explosions near enough to trigger mass extinctions were very uncommon, on the scale of one per few billion years. By using supercomputer models to determine the estimated masses of the dying stars and the intricate paths these radioactive materials followed, German astrophysicist Dieter Breitschwerdt and his colleagues found the most likely timings and locations of the explosions: two supernovae between around 290 and 325 light-years from the Sun.[1]

Despite the likelihood that the supernovae analysed by Wallner and his colleagues would not have been close enough to Earth to cause mass extinctions, he hypothesised that the explosions' radiation would have affected Earth's temperature. He said that the earlier event they uncovered happened at the same time as temperature shifts about eight million years ago, during the late Miocene period, when flora and animals all around the world were shifting. On top of that, the more recent incident occurred at the same time that Earth started turning colder, towards the tail end of the Pliocene period about three million years ago—a climatic change that may have aided in the birth of the human lineage.

Yet, in addition, Breitschwerdt pointed out that the radiation released by these explosions might have caused mutations in Earth's life forms. According to Breitschwerdt, "it would be feasible that an increased rate of mutations directly affected evolution, including the enlargement of the human brain."

In addition to toxins and radiation, epigenetic changes can also cause mutations. Epigenetic changes in the DNA activate or block certain genes without affecting the DNA itself. Consider it as if certain switches are turned either on or off.

Epigenetics can explain phenomena that biology cannot explain, such as discrepancies between genetically identical twins. Certain environmental or lifestyle variables have been shown to cause epigenetic modifications that can be handed down from generation to generation. Because gene mutation occurs at a far slower rate, epigenetic modification explains why organisms may develop and adapt so swiftly every generation.

The Dutch Hunger Winter of 1944 is a cruel but clear illustration of how sudden epigenetic change may occur.[3] The trains came to a standstill in the Netherlands in September 1944. Dutch railway employees thought that a strike would halt Nazi troop transit and aid the approaching Allied army. However, the Allied effort failed, and the Nazis punished the Dutch people by cutting off food supplies, resulting in famine (More than 20,000 people had perished of starvation when the Netherlands was freed in May 1945).

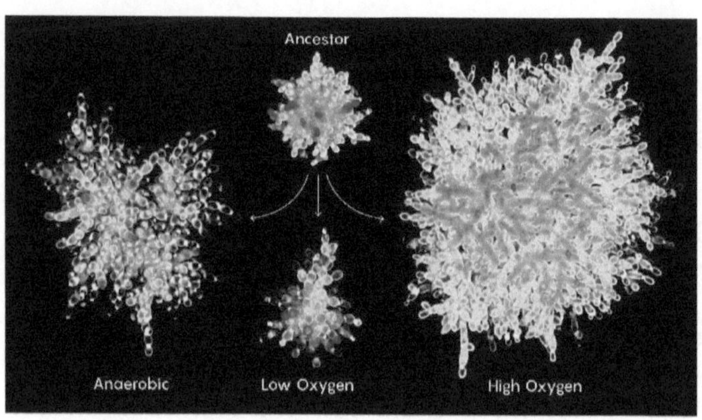

7) Under different environmental conditions, snowflake yeast evolve into significantly different forms. The ancestral form is shown at top. At bottom are the forms that evolved under anaerobic, low-oxygen and high-oxygen conditions. These epigenetic alterations are caused by dramatic changes in environmental conditions.

The Dutch Hunger Winter served as an unintended experiment in human health since it began and ended so quickly. Pregnant women were especially vulnerable, and the children they carried suffered from starvation for the rest of their lives.

The "Hunger Winter Babies" were a little bit heavier than normal when they reached adulthood. They had greater triglyceride and LDL cholesterol levels in middle age, as well as increased risks of obesity, diabetes, and schizophrenia. Those dangers had taken a toll on them by the time they reached old age. People born during the famine perished at a 10% greater rate than those born earlier or later. The Dutch Hunger Winter is only one of many examples of sudden epigenetic alterations caused by dramatic changes in environmental or lifestyle conditions.

These ranges are often significantly less for varieties. Similar to how variations tend to have smaller distributions than their parent species, closely related species tend to have a smaller habitat. Because they are indistinguishable from species unless, first, intermediate connecting forms are discovered, and, second, a sufficient level of diversity is present, varieties have the same general characteristics as species. Classifying a species is challenging when it has so many subtle yet distinct variations. In excellent studies, many insects found on the small Madeira islands are described as varieties, but there's no question that many entomologists would classify them as separate species. Some zoologists still classify some Irish creatures as species, even though they are usually considered to be variations.

Consider the classic case of *Primula veris* and *Primula elatior*, also known as the primrose and cowslip. These plants have wildly varying outward appearances; they each have their own distinct flavour and aroma; they bloom at slightly varying times; they prefer slightly different soil types; they climb mountains to varying altitudes; they inhabit varying regions; and, according to the results of a large number of experiments conducted over the course of several years, they are very difficult to cross. This is unquestionably the strongest evidence to date that the two varieties are unique from one another. However, they are connected by several intermediary links, and it is very debatable whether these linkages are hybrids. Therefore, as far as science can tell, there is an overwhelming quantity of experimental data suggesting that they descend from common parents, and thus must be listed as varieties. Fortunately, in most cases, closer examination will lead naturalists to a consensus on how to rank questionable varieties.

It is interesting to note that the most successful and dominating species of the larger genera tend to vary the most, and that variations often evolve into new and different species. For example, in areas where there are a lot more species than usual, the average number of variants per species is higher than in countries where the average is lower. Somewhat more of the most widespread and much spread or dominant species will be found on the side of the bigger genera than those in the smaller genera being placed on the other side. As a result, the more numerous and diverse the descendants of the most successful species, the more likely it is that the species will continue to dominate its habitat. However, via a series of events, the bigger genera tend to split up into smaller genera as well. And so it goes throughout the world, with all forms of life eventually stratifying into hierarchies of subgroups.

The nature of the many other connections of growth is completely beyond our comprehension, but in general, alterations to the structure of an organism at an early age will have an effect on later-developed components.

Darwinism

However, all of these processes are based on random chance. But how can so much be down to chance when all living creatures adapt perfectly to their environment?

The earliest sparks of scientific understanding of evolution, developed in the mid-19th century, clarified this intriguing question. Before that, most Western scientists held the notion that God created humans in his own image. They believed that all of the planet's species were created by a divine force. However, everything changed when Charles Darwin discovered something remarkable. Some experts had already discussed species evolution, but the British scientist was the first to provide clear evidence for how evolution may occur: through natural selection. His idea drastically altered biology, providing a new account of the genesis of humans, and established him as one of history's most noteworthy scientists and thinkers. But he had to take an exceptional journey and spend 20 years developing his ideas to achieve his goal.

Charles Robert Darwin was born on February 12, 1809, in the peaceful English Midlands market town of Shrewsbury. His father was a prosperous and well-regarded physician, and Darwin's mother, who died when he was only eight, was the daughter of Josiah Wedgwood, of pottery fame.[0]

In 1831, Darwin was 22 years old and studying at the University of Cambridge when he was asked to join a great expedition as a naturalist. The trip, led by Robert Fitzroy, set sail from England to examine the shores of South America. Darwin boarded the HMS Beagle and spent nearly five years journeying across multiple continents, beginning in South America, bringing back dozens of living specimens, pictures, and fossils. While on the HMS Beagle from 1831 to 1836, Darwin led a full and happy life, filled with fascinating adventures and specimens to gather to solidify his name and keep him busy for decades. Among his many achievements and experiences is the discovery of a new species of dolphin (which he dutifully named *Delphinus fitzroyi*), his survival through a devastating earthquake in Chile, his diligent and useful geological investigations across the Andes, and his innovative and widely praised theory for the formation of coral atolls (which, not coincidentally, suggested that atolls could not form in less than a million years). After being away from home for five years and two days, he returned to his family's residence in 1836 at the age of 27. As far as we know, he never left England again.

These fossils provided him with one of the first hints about evolution. He concluded that the similarities between the bones of a milodon (a colossal animal akin to the sloth), a gomphothere, and the remnants of an extinct horse were most likely not a coincidence and that there had to be some form of connection. When Darwin arrived in the Galapagos Islands, he also saw its gigantic tortoises (member of the turtle family), which resided on neighbouring islands but had distinct physical traits. Darwin noticed that the tortoises in humid areas with plentiful plants and vegetation had short necks and dome-shaped shells, while on the island with a drier environment, they had saddle-like shells and longer necks.

Additionally, after his return, Darwin spent a lot of time examining how animal breeders and guardians crossbred animals of different types to generate new variations. Man's artificial selection was important to the achievement of that creation. Darwin recognised that the natural world most likely used the same type of selection. He still felt the need to explain how this process took place with factual evidence.

Until he read the work of Thomas Robert Malthus, an 18th-century British philosopher, he got a broader picture. In an article, Malthus claimed that Europe's population would outnumber the food supply, resulting in "a fight for survival." This concept aided Darwin in explaining how evolution works: in nature, there is a battle for survival, and the strongest creature is not always the survivor. Instead, it is the one that best adapts to its environment. If a living organism possesses any characteristic that aids in survival, it will be more successful in reproduction. Those who do not adapt will die with less or without offspring. The organisms that have the most success in reproducing pass on their qualities to their offspring, and so on, until these variations become a new species.

As a result, the differences between Galapagos tortoises resulted from evolution. Those with longer necks could readily reach the plants to acquire food in a drier environment, but those with shorter necks and dome-shaped shells could consume grass and shelter themselves from predators in a humid environment.

A flock of finches (a tiny to medium-sized passerine bird) living on the Galapagos Islands is another outstanding illustration of this. These birds, known as Darwin finches, are well-known in science. According to legends, a flock of finches was blown off their usual course hundreds of years ago, probably due to a big storm.

As a result, the finches found themselves in a radically different habitat when they arrived in the Galapagos Islands in the middle of the Pacific Ocean. Luckily for them, it was a finch's paradise, with plenty of food and no natural predators. After a peaceful period of fast reproduction, there were eventually too many finches, leading to famine. Due to this, the finches became each other's competitors.

Because of their distinctiveness and tiny distinctions, some of the birds were able to avoid competing with their

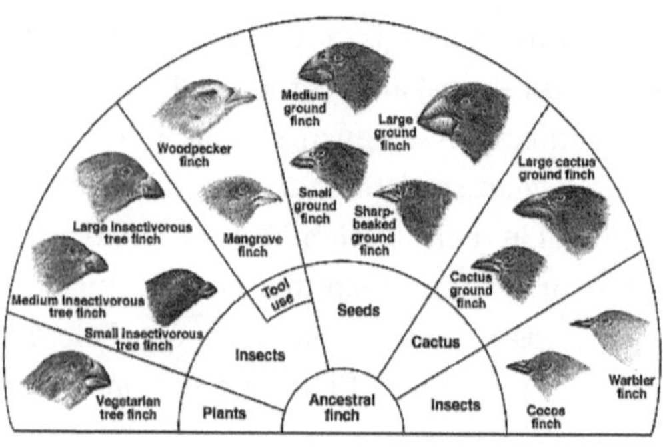

8) An overview of the different finches and their ecological niches.

fellow finches. Some finches gradually evolved superior beaks for digging for worms, whereas others evolved better beaks for breaking nuts. As a result, the finches sorted out ecological niches where they were secure from fierce competition. It was indeed "survival of the fittest." These characteristics were refined through many generations, allowing the finches to utilise their niches successfully. The differences between the worm diggers and the seed-crackers became so significant that they eventually could no longer produce fertile offspring with each other, resulting in the emergence of new species. There are about 14 various types of finch species surviving on the Galapagos Islands today, and all descended from one group of stranded finches.

Darwin had written reams following his journey, but none had been published. He needed their rebuttable proof, maybe because he anticipated his hypothesis would raise quite a controversy. But all changed when Darwin got a letter from Alfred Russel Wallace, a fellow naturalist and admirer, informing him that he had come to the same conclusion that evolution was caused by natural selection. Darwin became anxious that Wallace may claim full credit for the hypothesis, so the two naturalists agreed to present a joint letter detailing their observations. However, Darwin published his book *On the Origin of Species* a year later, and he became a celebrity far beyond the scientific community. His findings shook the foundations of the modern world. Yet, there is no mention of "survival of the fittest" in any of Darwin's works (though he did express his admiration for it). A full five years after the 1864 publication of *On the Origin of Species*, in Herbert Spencer's *Principles of Biology*, this term made its first appearance.

Darwin's book was named *On the Origin of Species*,[2] but he ironically failed to address the question of its title's namesake. Darwin's theory outlined a process by which an existing species may improve its survival by adapting to its environment and becoming stronger, better, and quicker, but it offered no clues as to how it might generate an entirely new species.

Fleming Jenkin, a Scottish engineer, thought about the issue and found a significant fault in Darwin's theory. If a desirable characteristic emerged in one generation, Darwin reasoned, it would be handed on to the next, making the species stronger in the long run. However, Jenkin made the argument that a positive characteristic in one parent would be diluted by mating in subsequent generations rather than becoming dominant. Whiskey diluted with water in a tumbler is weaker whiskey, not stronger. The already weak solution will get even weaker if you pour it into a second glass of water. Any beneficial characteristic passed on from a single set of parents would be gradually diluted by further matings until it was no longer present. As a result, Darwin's hypothesis was not a formula for flux but rather for stability.

A lucky fluke would occur sometimes, but it would quickly disappear as the main tendency was to restore things to a steady mediocrity. Some hitherto unanticipated mechanism was needed if natural selection was to be effective. It was a retiring monk named Gregor Mendel, 800 kilometres away in a peaceful nook of Middle Europe, who came up with the solution, unnoticed by Darwin and everyone else. Mendel studied Darwin with great interest, but he disagreed with the blending hypothesis. Instead, he hypothesised that traits, such as eye colour, height, and flower colour, were conveyed by small particles that were inherited whole from one generation to the next. The patient monk bred and crossed pea plants to determine how a few specific qualities, like height, were transmitted. When Mendel bred a tall plant with a short plant, the resulting offspring were always tall and never of a medium height. When he then bred these offspring together, he produced three tall offspring and one short offspring. Mendel understood precisely what this meant. Height was transmitted via what we now call a gene (though Mendel never used that term himself). However, no one understood the significance of Mendel's findings at the time. Around 1900, his work was rediscovered, and he was given credit for one of the greatest discoveries in the history of science.

Darwin and Mendel unwittingly created the basis for all of 20th-century biology. Mendel's work offered a method to explain how this might happen, but Darwin's insight that all living things are related, that ultimately they "trace their genealogy to a single, common source," was crucial.

The Universal Struggle

How have such precise adjustments been made from one section of the system to another, from one distinct organic creature to another, and from the conditions of existence? These lovely adjustments are most obvious in the woodpecker and the mistletoe, but they are also present in the smallest parasite that attaches to the hairs of a quadruped or the feathers of a bird, in the body of the beetle that dives through the water, and in the feathered seed that floats on the lightest breeze.

Due to this competition for survival, any variation (no matter how small and simple or how it may have arisen) that proves beneficial to an individual of any species' survival will generally be passed on to its offspring. As a result, the next generation will have a greater chance of survival, which is because only a small fraction of the many individuals of each species who are born regularly will actually make it.

Parent and offspring morphological structures change throughout time as a result of natural selection. In social animals, it will modify everyone's anatomy in such a way that everyone benefits from the chosen transformation. What natural selection can't do is alter the structure of one species without benefiting that species in any way for the benefit of another.

In order to clear themselves of whatever is harmful to their fluid, certain plants secrete a sweet juice. Insects will travel considerable lengths to obtain this juice. In this situation, pollen would be spread from one bloom to another by insects that visited both flowers for their nectar. We have excellent cause to expect that the process of crossing the blossoms of two different individuals of the same species would create exceptionally vigorous seedlings, which would therefore have the highest chance of prospering and surviving. It's likely that some of these offspring will develop into full-fledged nectar-producing adults. Ultimately, success came to the flowers that had the biggest glands or nectaries and secreted the most nectar since they were the most frequently visited and crossed by insects. A flower would be picked if its stamens and pistils were arranged in a way that made it easier for pollen to be transferred from one blossom to another, taking into account the size, shape and behaviour of the specific insects that visited it. Pollen is made for the sole purpose of fertilisation, so its "loss" is negligible to the plant.

Yet, since distinctive features often arise during domestication in one sex and become fixed to that sex, the same phenomenon likely occurs in nature. This is dependent not on a battle for survival but on a battle for reproductive dominance between males; the losing male will not starve to death but will produce less (if any) offspring as a consequence of the competition. It follows that sexual selection is not as stringent as natural selection.[2]

It has been reported that male alligators battle over females, that male salmon fight all day, and that male stag beetles frequently show scars from the giant mandibles of other males. Males of polygamous animals—mating with multiple partners—appear to be the most heavily armed, suggesting that they wage the fiercest wars. In the bird world, the competition is usually of a calmer variety: the males of many bird species engage in fierce rivalry in an effort to attract the females with their songs or plumage.

Natural selection relies in large part on the fact that organisms are sometimes left alone for long periods of time. Unless the area is extremely vast, the natural and inorganic circumstances of life tend to be quite consistent in a limited or isolated area. If the size of a remote place is small because it is surrounded by obstacles or because of very unusual physical conditions, then the number of individuals supported there will also be small. Fewer individuals mean fewer opportunities for favourable variations to arise, which slows the process of natural selection and, thus, the emergence of new species. A smaller area means less drastic adaptation and extinction rates due to the competition for survival.

To better understand this, let's examine the chain of tepuis in South America. Before 200 million years ago, the Guiana Shield, a collection of ancient pre-Cambrian continental crust, was once connected to Africa. The mystical tepuis dominate the horizon; they are lofty, cloud-covered tabletop mountains that rise hundreds of metres from the lush flora on sheer sandstone cliffs. In this region of South America, there are hundreds of tepuis, many of which have never been visited by humans. Their summits exist alone as islands, each harbouring a distinct ecosystem that has developed uncontaminated by the genetic makeup of its neighbour, separated by an impassable gulf (only birds and their droppings visit these heavenly islands).

One of these places is the Auyán-tepui mountain in Canaima National Park, a UNESCO World Heritage site in the Gran Sabana area of Bolvar State, which features an outstanding waterfall. The Angel Falls is the tallest uninterrupted waterfall in the world, with a height of 979 metres and a drop of 807 metres!

9) The Auyán Tepui, located in the Bolívar state of Venezuela, has a summit area of 666.9 km^2 and is a popular destination for tourists. It is one of the largest tepuis in the Guiana Highlands, although it is not the highest.

Part of what makes the biology of these tepuis so fascinating is that the plants and creatures on the summits of some tepuis are more closely linked to those in Africa than to those found in the jungle below them. This is (besides the fact that small islands have less drastic adaptation and extinction rates due to the competition for survival) because the summits have been separated from the jungle below them for such a prolonged period of time that they more closely resemble the environments of the regions they originally connected. Only until the supercontinent Pangaea began to break apart some 200 million years ago did these tepuis become separated.

But coming back to our story, with the help of the following example, let's endeavour to gain a greater grasp of the pace of the intricate evolution of new species.[2] Let A–L stand in for the human species of the genus *Homo,* which is in huge numbers in its habitat; these species are assumed to match each other in varying degrees.

Assume that (A) is a widely distributed and diverse species of a large genus within its habitat. It is expected that the changes will be small but exceedingly varied, that they will not all show at once but instead occur at varying intervals, and that their durations will vary widely. Those variants will only be maintained or selected by natural means if they are beneficial in some manner. Here, the concept of advantage obtained from divergence of character comes into play, as it is this principle that often causes the most distinct or divergent variants to be retained and collected through natural selection.

Two quite distinct human variations, a1 and m1, are predicted to have emerged from species (A) after approximately 1,000 generations. Since the propensity to variation is inherently heritable and these two types will continue to be subjected to the same situations that caused their parents to be variable, they will also tend to fluctuate, and typically in roughly the very same way that their parents changed. In addition, these two variants, being only slightly tweaked forms, will strive to pass down those advantages which made their lineage (A) more numerous than most of the other inhabitants from the same region.

If these two types are variable, then the most extreme differences between them will likely persist for another 1,000 years. And after this time period, it is assumed that variety a1 gave rise to variety a2, which would diverge further from (A) than did variety a1 according to the concept of diverging. After 1,000 generations, some of the variations will generate only one variety, although in a greater and more modified form, while others will generate two or three, and others won't produce any at all.

It is assumed that after 10,000 generations, species (A) will have given rise to three forms, a10, f10, and m10, which will have diverged in personality during the successive generations and will now differ substantially from each other and from their common parent, though this divergence may be unequal. Eight species, denoted by the letters between a14 and m14, are the result of carrying out the same process for a larger number of generations (A). Six new species, designated n14 through z14, are expected to emerge after 14,000 generations. If you look at each genus, the species that are the most distinct in personality are the ones that tend to generate the most offspring. This is because they are the ones most likely to adapt to new and diverse niches in the natural order.

Natural Selection

Yet, some physical adaptations are relatively easy to spot. Bumblebees' long tongues are well-suited for reaching into the deep crevices of the corolla to scoop up nectar. Short-tongued bees, sometimes known as "nectar thieves," circumvent this challenge by chewing a hole in a petal just below the corolla. Most birds' wings are similarly optimised for flight manoeuvres, including flapping, swooping, and soaring. It's not that birds that received genes that coded for more effective wings were at an evolutionary advantage, but rather that they were more likely to survive and reproduce.

The tongue of a bumblebee or a bird's wings is each adept at carrying out its allotted tasks. All around us in the natural world are examples of the beauty, symmetry, and effectiveness of biological co-adaptations. Although Darwin did not use the word "coevolution", he suggested how plants and insects could evolve through reciprocal evolutionary changes.

However, evolution is not without flaws. It would be a mistake to believe that problems always have the simplest and most cost-effective solutions, as that is not how natural selection works. Some adaptations appear to have been designed by a skilled and inventive biological engineer, whereas others appear clumsy or poorly thought out.

One really good example is our own eyes. It's true that human eyesight is exceptionally sharp and vivid. If they are in top condition, they will focus quickly and auto-expose correctly. Also, they have an innovative design that incorporates a self-cleaning cavity. Nevertheless, many of us have poor eyesight, and cataracts are frequent. The light-sensitive retina is located behind a layer of blood vessels and nerves, which significantly reduces the quantity of light that can reach the retina (therefore, it is a major "design" fault). A hole must be present in the retina for the blood vessels and neurons to reach the brain in this configuration (the blind spot). More significantly, it indicates that retinal detachment is a common occurrence. Large cephalopods like the squid and octopus exhibit the superior "design" of positioning the retina in front of the eye.

Evolution's flaws may be traced back to the constant need for adjustment. Originally, the human eye was just a small area of light-sensitive skin cells. After the fundamental anatomical components of the eye have been established, such as the connection of light-sensitive cells to the brain through nerves, evolution cannot go back and entirely restructure the system. The backwards changes nearly never occur.

The fact that natural selection incorporates several different types of evolutionary forces at once just adds to the complexity. There is no one-and-done way to improve our eyesight that does not have unexpected consequences elsewhere. It's possible that the visual processing area of the brain utilises resources that may be put to better use elsewhere. We are full of make-do and compromises since no adaptation happens in a vacuum.

Humans' small inner ear bones, the malleus and incus, or "hammer" and "anvil," are excellent examples of this phenomenon since they transmit and amplify sound vibrations on their route to the ear drum. The tiny bones in your ear and the reptiles' lower jaws are quite similar. Various commonalities exist because these bone types are inherited from a common ancestor across mammals, contemporary reptiles, and even fish. The distinction is that mammals adopted these skeletons during evolution. Because of their previous function, the adaptation is constrained; the transition from the jaw bone to the ear bone could only occur via mechanisms that provided a selective advantage for the species in question. The optimal solution cannot be arbitrarily selected by natural selection and implemented immediately. A new human ear design wouldn't have to rely on any antiquated components, which would make it superior to any previous attempts at creating such an organ.

For the time being, nevertheless, it's crucial to stress that just because the human brain seems tailored for a certain activity doesn't indicate that it actually is. Determining whether or not something is an adaptation is somewhat simpler in animals, especially those whose behaviour appears to consist entirely of innate reactions and processes. The ability to navigate the heavens by birds and the organisation of social insects into distinct castes of workers both appear to be the product of natural selection and so encoded in the DNA of the organisms involved. They're both effective answers to exceedingly difficult challenges that organisms have faced throughout their development. If we take reading as an example, the likelihood of either of these species learning to read from scratch is low. An equally improbable explanation is that they emerged as a result of genetic drift, or a succession of random mutations.

Due to evolutionary pressures, a bear that lacks one or more favourable features or physical aspects will perpetuate a lineage of offspring with their own less successful traits. We briefly mentioned the fact that this is not the consensus, but natural selection is more likely to put an end to the lineage of "less-successful" bears before the lineage of bears whose ancestors passed on "more-successful" features.

It's also possible to flip the perspective and look at it from the opposite side. Over several generations of bears, newly evolved "successful" characteristics and physical attributes may arise as a result of these evolutionary pressures.

For instance, evolutionary forces might cause a larger and stronger bear to appear after many generations. In this scenario, the "new" trait increases the likelihood of survival through nature's selection process. At first look, the difference between the somewhat larger and stronger bear and its smaller and weaker fellow bears may seem insignificant. However, the progeny of the larger and stronger bear will continue to grow larger and more robust. The larger and

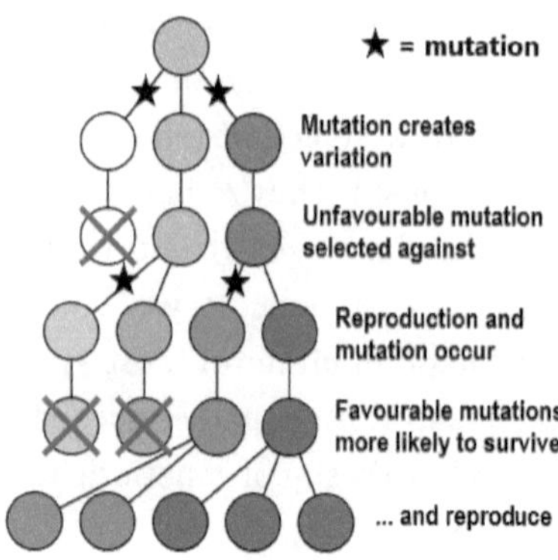

10) The process of mutation followed by natural selection leads to the emergence of a population with distinct traits compared to their predecessors, such as alterations in skin pigmentation, levels of aggression, intelligence, metabolic rates, body morphology, etc.

stronger bears may ultimately diverge enough from the smaller and weaker bears that they are unable to create fertile offspring with each other, just as we witnessed with the Darwin finch story (However, we're talking about hundreds, if not thousands of generations here).

What is considered to be a desirable trait, or what characteristics endure the tests of nature's selection process, is entirely environment-dependent. On an island with a scarcity of food, smaller, less energy-consuming, or better fat-retaining bears will thrive better because natural selection favours them for being specialists in securing "safe" ecological niches. For a moment, let's assume that the reduction in size to conserve energy, be less energy-demanding, and better fat-retaining will provide a substantial advantage and increase chances of survival. If this is the case, the bears' ability to successfully implement their modifications will be honed over many generations. The smaller, less energy-demanding and better fat-retaining features may never have evolved if the same island had different circumstances, such as plenty of food.

Perhaps a better and "real-life" example of these furry animals is the evolution of the polar bear. Originally a subspecies of the brown bear, the polar bear adapted to its Arctic habitat by turning its fur white. New evidence gathered by an international research team shows how soon this new species diverged from the brown bear. When scientists studied the genetic makeup of brown and white bears, they found some interesting differences. Most importantly, the polar bear just emerged as a distinct species over the last few hundred thousand years. According to Professor Rune Dietz of the Arctic Research Centre in the Department of Bioscience at Aarhus University, this suggests that the region is far younger than was previously believed.

The researchers hypothesise that brown bears migrated northward during a warmer climatic era and that when a colder time eventually arrived, a subpopulation of brown bears may have stayed isolated and been forced to adapt rapidly to the new colder environment. Natural selection has altered the genes that regulate blood-fat transport and fat breakdown in the same way as mutations in these genes have caused the fur colour to change from brown to white. Asides from the colour, over the course of hundreds or thousands of generations, the differences between the brown and polar bear became so great that they eventually were unable to produce fertile offspring with one other, leading to the formation of new species: brown bear (*Ursus arctos*) and the polar bear (*Ursus maritimus*).

Contrary to polar bears, brown bears prefer to spend their days alone. They are quite competitive with one another in the mountains. They seldom get together in more than a group of three. They are eating machines; therefore, they don't bother with table manners. They live by the motto, "Mine, mine, mine!" The only real exception to this rule is when salmon swim through rivers. Even yet, brown bears in the north seldom attend these types of gatherings, but Kodiak bears in western Alaska frequently do. Sharing food is more commonplace among polar bears, especially in big groups. Although polar bears are not known for being very sociable, whenever they find a dead whale, there is usually enough food for everyone, so there is typically no fighting (Tolerance has been shown between members of groups as large as twenty to thirty). They have reached an agreement with one another. Again, this demonstrates how unprepared humans are to deal with an animal that is not willing to share, even if they are physically capable of doing so. Because of this, not only do these mighty predators differ from one another, but the ecological society itself also shifts as a result of adaptation to the environment.

Both predators and their prey undergo change as a result of their interactions with one another, with the predator becoming more efficient hunters and the prey becoming more particularly skilled at evading them. Natural selection acts on both entities simultaneously due to their coevolution. As a result, prey and predators typically engage in an evolutionary arms race, with the prey developing strategies to counteract the predator. The same holds true for herbivores and the plants they consume.

The development of mutually beneficial relationships between plant species and the insects and birds that pollinate them is another example of this. In this regard, flowers and their pollinators have evolved mutually beneficial adaptations: flowers display their beauty to attract pollinators, while insects and birds have adapted to feed on the nectar and pollen they find there.

Flowers and bees have coevolved together in ways that most individuals are unaware of. Around 120 million years ago, some wasps gave up eating meat and learned that pollen and nectar were good foods to eat, eventually evolving into a new species: bees.

Meanwhile, plants and bees (co)evolved at the same time. As plants learned that it was good for them to be able to attract these bees, they became more colourful and showy in the middle of the bees' visual range. Another example is that the snapdragon only wants to attract bumblebees, so it gives off four times more scent during the day, when bumblebees are active, than at night.

Another form of pollinator that has coevolved for mutual benefit is the hummingbird. The nectar from the flowers provides essential nutrients for the hummingbirds, which in turn help to pollinate the flowers. Flowers of a certain hue attract hummingbirds, while others bloom just in time to suit the birds' beaks during mating season. Flowering plants have coevolved with different kinds of hummingbirds, as evidenced by the unique form and length of the corolla tubes, which have adapted to the shape and length of the hummingbird beak that pollinates that plant. While the bird is feeding on the nectar, the pollen will stick to a specific section of its body due to the flower's altered form.

Coevolution between humans and plants can also be seen, which is an intriguing phenomenon in its own right. Just as plants have evolved to produce toxins to protect themselves from being eaten, humans have evolved methods of avoiding them. Bad smells and tastes often indicate the presence of harmful compounds. If we inadvertently eat something toxic, we have secondary lines of defence: gagging, vomiting or diarrhoea. If you are a woman who has been pregnant, you'll probably have experienced for yourself some of our ancestors' ancient survival skills when it comes to food.

A fetus's major organs develop between six and fourteen weeks after conception. At around the same time, the mother's immune system temporarily lets down its guard, giving the fetus free rein to establish itself in the uterus. As a result, pregnant women are particularly susceptible to bacteria and viruses during those weeks. Less than absolutely fresh meat, for instance, may well contain organisms such as toxoplasma, a type of parasite. Normally, our bodies would destroy such pathogens without a problem, but in early pregnancy, they can cause maternal infection and even trigger a miscarriage. Over thousands of generations, it appears humans have evolved a strategy for protecting the developing fetus against dangerous poisons, which really seems to be an evolutionary explanation for morning sickness.

Given the methods and forces mentioned for biological evolution in nature to evolve into an unbounded number of species, we may infer that Charles Darwin revolutionised the game by explaining that variety emerges from a biological process. Darwin categorised humans at the exact same evolutionary stage as all other species on the planet. Scientific developments have verified his idea, and even the Catholic Church has come to acknowledge that evolution is compatible with faith decades later.

Convergent Evolution

Before moving on to the diffusion of life and ultimately the rise of the human species, we must first understand the last fundamental principle of nature's evolutionary process: convergent evolution. When species occupy similar ecological niches and adapt in similar ways in response to similar selection forces, convergence evolution occurs.

One of the most well-known examples of convergent evolution is the evolution of comparable structures in species that lived at separate times. The eye of cephalopods (such as squid and octopus), vertebrates (including mammals) and cnidaria (such as jellyfish) are perfect examples. A variety of processes led to the progressive refinement of eyes from the simple photoreceptive spot of their last common ancestor. However, the cephalopod eye is "wired" in the opposite direction, with blood and nerve vessels entering from the back of the retina, as opposed to the front as in vertebrates, resulting in the absence of a blind spot. Convergent evolution can also be shown in structures other than eyes.

Many different kinds of fish, marine mammals like dolphins, and even Mesozoic fish-eating reptiles called ichthyosaurs all converged on a single, streamlined form. *Phylliroe* (a type of sea slug) and other snails share this body structure and these swimming adaptations. Some aquatic animals have adapted a fusiform body shape (a tube tapering at both ends) so that they can swim quickly despite the considerable resistance they encounter. Both earless seals and eared seals have bodies that are similar in shape: they both have four legs, but those legs are highly adapted for swimming. Despite evolving in separate, independent clades, the marsupial fauna of Australia and the placental mammals of the Old World share a number of startlingly comparable characteristics. The thylacine's (Tasmanian tiger's or Tasmanian wolf's) physique and, more specifically, its skull form, converged with those of Canidae, such as the red fox (*Vulpes vulpes*).

Although they are not related, bats and birds both eventually evolved and developed wings capable of flying. Because both birds and bats evolved from terrestrial tetrapods, their limbs are homologous, but their flying mechanisms are just similar. This makes their wings an example of functional convergence. Each subgroup has evolved its own method of powered flight.

However, there is a major design distinction between their wings. When compared to the bird's feathered wing, which is securely linked to the forearm (the ulna) and the strongly fused bones of the wrist and hand (the carpometacarpus), the bat's membrane wing is spread over four greatly lengthened fingers and the legs. However, whereas bat and bird wings share similar functions, they do not share similar anatomy.

On Earth, "flight" has developed four times in four distinct groups: birds, bats, pterosaurs, and insects. It's not because they developed on Earth that they all have wings; it was advantageous to fly, and wings are nearly the only method to fly. Consequently, we may anticipate that these limits operate everywhere in the Universe.

11) The wings of vertebrates exhibit partial homology, originating from their forelimbs, while serving as analogous flight organs in pterosaurs, bats, and birds, which evolved independently.

There are distinctions between bat and bee wings, although not in principle. Both are made comprised of a flexible membrane held together by a solid framework. Both methods provide lift by forcing air over the membrane. The primary distinction between bee and bat wings is not in their anatomy but in their function. Insects, due to their size, cannot rely on wing flapping like bats to achieve flight. In order to fly, they must buzz, creating lift on both the forward and backward strokes of their wings. We can observe how tightly limited these solutions truly are, so we can be more sure in our forecasts despite the diversity of implementations on our globe. Although birds, bats, and bees all have unique wing structures, they have all ultimately developed wings that are aerodynamic despite the vastly varied physical restrictions under which they have evolved.

And while we're on the subject of bats, it's worth noting that bats and cetaceans (whales and dolphins) independently developed the sensory adaption of echolocation from the same genetic changes. Passive electroreception was independently established by the Gymnotiformes of South America and the Mormyridae of Africa (about 119 and 110 million years ago). Both groups developed active electrogenesis some 20 million years after that, creating weak electric fields to help in hunting.

Blue eyes and light skin are examples of convergent evolution in humans. As we shall see in the latter chapter of the next part (regarding the out-of-Africa migration movement), humans eventually made their way to cooler climates further north. Since sunlight is less in the northern latitudes, humans benefit from lighter skin. A lack of melanin (a substance in your body that produces hair, eye and skin pigmentation) in the skin allows for greater "exposure" to the Sun, which is especially essential in the northern latitudes where vitamin D synthesis requires the conversion of cholesterol in the skin's cells to energy. Some skin-lightening genetic changes were shared by both Europeans and East Asians, suggesting that there was at least some discolouration before the two groups parted. The genetic changes that reduced the pigmentation density of the skin in both groups occurred when the lineages separated and became genetically isolated.

This is an example of convergent evolution, which occurred because some selective pressures are ubiquitous. Like a plant whose roots are bending around rocks in search of softer and richer soil, all living things have an innate need to obtain the nutrients they need to thrive and reproduce.

Once again, nature has a knack for devising ingenious means of acquiring what it needs. Therefore, having eyes is a good thing because of how common light is. Nature tends to reuse successful ideas as they arise.

Due to the universality of nature's selection pressures, this may suggest that evolution follows a similar path on planets with similar circumstances to Earth. As an example, for the sake of the remaining bit of this chapter, let's hypothetically construct a fictional scenario where the origin and development of life is a universal principle, where, given the right conditions, extraterrestrial life can evolve to the complexity of *Homo sapiens*.

If this is actually the case, our fictional complex extraterrestrial life might not mirror life on Earth exactly, but it might have some similar biological mechanisms that are recognisable to us. It's theoretically possible, according to the theory of convergent evolution. If it worked for life on Earth, it might work everywhere where the conditions are right.

If we are discussing the same complexity as *Homo sapiens* and having the power to leave the biosphere, we can quickly hypothesise one identifiable mechanism by which analogous extraterrestrial life should have evolved: it would need to know how to use tools (Everything you see around you, from the book you're holding in your hands to the light that allows you to read it, as well as clothing, food, transportation, and spacecraft, is the result of the use of tools). Therefore, digits, such as fingers or toes, or something somewhat equivalent would most likely be a part of an extraterrestrial species' anatomy. It would also be beneficial and conceivable for large heads to house complex brains, allowing for ever greater levels of intelligence and the solving of increasingly difficult problems.

Tens of billions of neurons make up the human brain, making it incredibly intricate and sophisticated. It's not completely out of the question that similarly intelligent extraterrestrial species might likewise have a (relatively) large, sophisticated cognitive organ. As a result of our intelligence, we no longer needed fur to keep us warm because we could make clothing for ourselves, suggesting that similarly advanced extraterrestrial species would not have fur or thick skin. Since we focused our creation on what was functional, it stands to reason that other sentient species in the Universe would have comparable or even identical physical characteristics.

Some characteristics are far more likely to be present than others for a variety of reasons. First, the high cost of pumping and speeding vast volumes of water makes the emergence of aquatic lifeforms that can dominate their environment and other planets less likely. Second, since the primary eye should always be turned in the direction you're moving or need to be aware of, it stands to reason that extraterrestrial eyes would evolve to face forward. Third, a complex cognitive organ (like a brain) is more likely to form at the front (on the primary eye side) since it is more efficient to have the neural system close to your sensors. This, in turn, makes having a head with eyes, ears, and a nose (primary senses) plausible configurations. Another one is that the head should be closer to the hands since the hands are near the senses. Furthermore, once you've reached a certain degree of complexity, you no longer value a generalist approach but rather one that emphasises specialisation. Perhaps there would also be a "practical" range of aggression that universally everybody shares.

On this planet, it is quite common for creatures to work together and exchange information. That seems strange at first look: after all, why would two separate organisms work together? But it doesn't take much reflection to see that they'll work together if they both benefit from it. If the rule is universal, it will apply in this case on any planet. It's possible that extraterrestrial species might work together for no other reason than the fact that they'd both benefit. However, a strong sense of relatedness fuels and motivates a lot of Earth's cooperation and sociality. Because of familial ties, species work together. We take care of our families by providing for our children, helping our parents, and fostering positive relationships among our relatives. Evolutionary theory provides solid support for all of these claims. Characteristics, features, and even habits may be traced back to our genetic make-up, which is handed down from parents to offspring. And if our genes predispose us toward helping our related siblings, we'll all prosper and increase the chances of passing on our genetic makeup.

The trouble is, since we don't know what extraterrestrials could possibly have for genes, it is impossible to execute this theory and form of logic properly. Therefore, we can't conclude that this is as universal as, say, the laws of gravity or the water cycle. The social organisation of extraterrestrial species may seem very different from our own if the way they relate to one another is radically different. The field of astrobiology often assumes that any extraterrestrial life would consist of carbon-based organisms. About 45–50% of Earth's dry biomass is composed of carbon, making it a fundamental element in all forms of life. Carbon molecules are particularly amenable to manipulation by enzymes due to their low mass and tiny size. Carbon atoms constitute the backbone of all biological molecules, and they are bound to other elements, such as oxygen and hydrogen, and sometimes nitrogen, phosphorus, and sulphur, to form complex biological structures (collectively known as CHNOPS). Carbon is unique among the elements in its potential to serve as a key support for biological systems and activities like metabolism. Silicon has been proposed as a possible replacement, as it is in the same group as carbon in the periodic table. Like carbon, it can create four valence bonds and readily attaches to itself. However, in silicon's instance, this self-adhesion takes the shape of crystal lattices rather than lengthy chains. In spite of these parallels, silicon is far more electropositive than carbon, and silicon compounds do not readily recombine into diverse permutations that may reasonably sustain life-like activities.

Composition isn't the only factor: selection forces also matter, as they drive adaptations in quite diverse directions. Plants and animals, for example, might reach heights much beyond what is possible on Earth if they lived on a planet with far less gravity. It would take less effort for their bodies to transport nutrients and blood (if extraterrestrials used those kinds of vascular systems). On the other hand, life on a planet with extremely high gravity would be stockier, shorter, and heavier. They'd require denser bones (or some other kind of support) to survive under higher gravity. There would be less light on a planet that is farther from its host star than Earth is, indicating that eyes would be larger on such worlds. The average temperature of a planet can have an effect on the size of its inhabitants. On Earth,

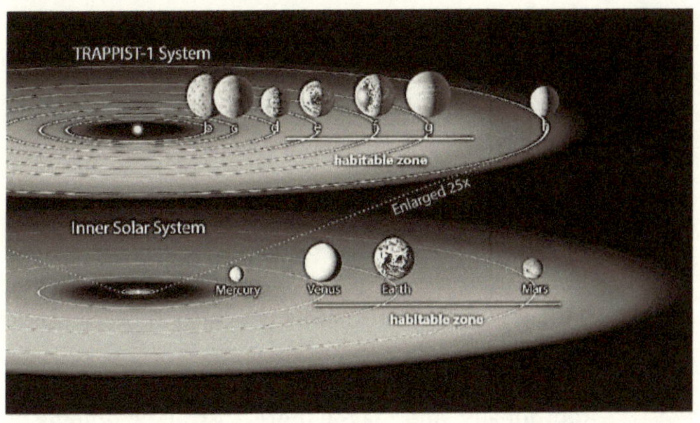

12) The TRAPPIST-1 system, one of the tens of billions of planetary systems in the Milky Way Galaxy (which in turn is one of the hundreds of billions of galaxies in the Universe), contains seven planets that are all roughly the size of the Earth. Three of them (TRAPPIST-1e, f, and g) inhabit the so-called habitable zone of their star. The habitable zone, also known as the Goldilocks zone, is a belt around every star where temperatures are just right for liquid water to exist on the surface of a planet. Nonetheless, if the evolutionary processes mentioned in this chapter are indeed universal, what are the probabilities and similarities of life in the TRAPPIST-1 system or elsewhere in the Universe?

"deep-sea gigantism" occurs because larger bodies can be more efficiently kept warm, but in deserts, tiny creatures have a greater mass-to-surface-area ratio, allowing them to dissipate heat more efficiently.

Details of animal form and appearance will forever be influenced by events of evolutionary (and maybe cosmic) history. For approximately 400 million years, a four-finned fish has been regarded as crawling out of the water and giving rise to our species, explaining why we have four limbs. If events in our evolutionary past had gone differently, we might have developed an extra pair of hands or feet. So it's unlikely that we'll ever find a truly close match with the corresponding species on another planet. However, some things are just so tightly constrained that there aren't really many alternative ways to do things.

This brief small detouring question of whether the genesis and development of life is a universal principle or not can teach us valuable things since astrobiology adds a multidisciplinary viewpoint to biology, which aids science in better understanding it.

Additionally, and maybe more abstractly, it helps us comprehend the origins and evolution of life, as well as the peculiarities and marvels of our existence and our place in the Universe. *Homo sapiens* learned that we are not at the centre of the Universe a little over 500 years ago. Not until the 20th century did scientists realise that the Universe was not static and had a beginning: the Big Bang. The billions of human beings living today all belong to one species: *Homo sapiens*. Humans, like all other animals, have a wide range of individual diversity in physical attributes, including height, weight, and even skin and eye colour. However, our similarities outweigh our differences by a wide margin. The idea that humans descended over millions of years from a common ancestor with other species is a relatively new thought. As we will see in the next chapter, through (perhaps universal) evolutionary pressures, the world is constantly changing. Sometimes radically, and we are evolving along with it.

3. THE ERUPTION OF LIFE

Due to the natural evolutionary processes and pressures discussed in the previous chapter, unicellular life eventually evolved into five kingdoms of nature, including the Monera, Protista, Fungi, Plant and Animal Kingdom. (The classification of nature's five kingdoms remains the most widely recognised today, despite recent advancements in genetic research suggesting fresh changes and reopening the discussion among specialists. Such is the case with Carl Woese and George Fox's sixth kingdom, which separated bacteria into the Archaea and Bacteria groups in 1977, and Cavalier Smith's seventh kingdom, which introduced a new category for algae titled Chromista to the previous six). However, as we shall see in this chapter, the intricate path from a primordial simple eukaryotic cell to the infinite biodiversity of this planet is an enthralling journey.

The Tree of Life

All current life on the planet may trace its genetic lineage back to a single ancestor, known as the Last Universal Common Ancestor (LUCA). The same genetic code is used by all biological species, including viruses, whose origins remain mysterious. In fact, the LUCA is not the first form of life on this planet but rather the most recent ancestor of all known forms of life. Based on a number of genetic studies, the currently recognised Tree of Life may be traced back to a split between the monophyletic Bacteria domain and the group formed by Archaea and Eukaryota. According to the most up-to-date

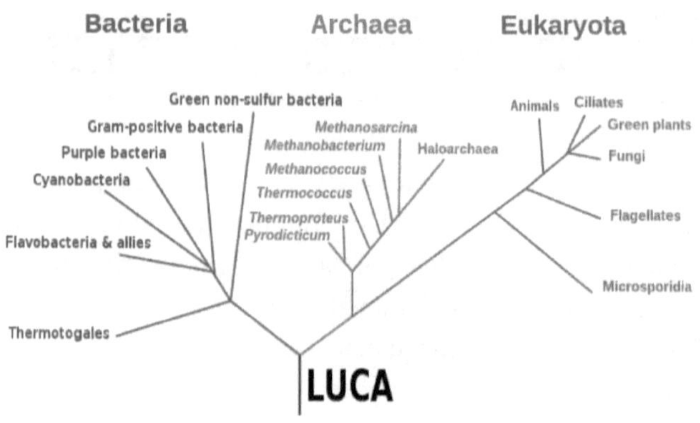

13) A 1990 phylogenetic tree linking all major groups of living organisms to the LUCA, based on ribosomal RNA sequence data.

genomic research, the Archaea and Bacteria domains make up the base of the evolutionary tree, with eukaryotes branching off from them. We humans and every other multicellular creature, including plants, fungi, and even single-celled protists, belong to the Eukarya kingdom because our cells have nuclei that separate our DNA from the rest of the cell.[3]

A single LUCA was strongly supported by the test, but this doesn't mean it existed in loneliness. Rather, it was the solitary cell whose progeny outcompeted the rest and made it through the Paleoarchean. The last universal common ancestor is the most recent population from which all organisms now living on Earth share common descent

There may not be any direct fossil evidence for the LUCA, but the fact that all known forms of life have so many molecular hallmarks suggests it must have existed. Genome similarities allow us to infer its properties. Transcription and translation systems, which transform DNA into RNA and then into proteins, are only two examples of the co-adapted properties described by these genes, which represent a complex living form. About four billion years ago, our LUCA most likely made its home in the hot water of deep marine vents close to bottom ocean lava flows.[3]

After the LUCA gradually developed into the mitochondrion, eventually evolving into a "fully nucleated" eukaryote, it ultimately developed an even more peculiar trick. It took a long time, perhaps a billion years, but when they perfected it, it was revolutionary. They learnt to join forces to construct complex multicellular organisms. This breakthrough made it possible for large, sophisticated, visible creatures like humans to exist. This amalgamation was truly a biological revolution. Planet Earth was set to go on its next big adventure.

The discovery of fossils is critical to our understanding of evolution and the ways in which plants and animals have adapted to their respective habitats. The discovery of fossils gives a record of how organisms evolved over time and demonstrates how this evolution may be shown as a "Tree of Life," which demonstrates that all species are connected to one another.[1] Yet, becoming a fossil, however, is not an easy task. More than 99.9% of all living organisms will perish in the process of decomposition, which will reduce them to nothingness. It is believed that only approximately one bone in a billion ever becomes fossilised, much alone gets excavated.

Trilobites are the group of extinct creatures with the greatest diversity that has been preserved in the fossil record. Near the beginning of the great outburst of complex life that

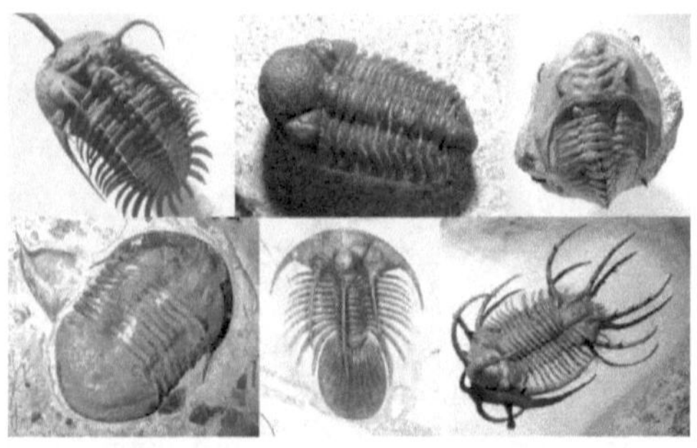

14) Montage of trilobite genera: Top row: *Walliserops*, *Phacops* and *Cambropallas*; Bottom row: *Isotelus*, *Kolihapeltis* and *Ceratarges*.

is popularly called the Cambrian explosion, approximately 540 million years ago, trilobites made their first appearance, fully developed and apparently completely out of nowhere. They then suddenly disappeared, along with a good deal of other life forms, during the great and yet still mysterious Permian extinction approximately 300,000 years later. Trilobites developed into a wide variety of ecological roles throughout this time period. Some trilobites travelled across the bottom as predators, scavengers, or filter feeders, while others swam and fed on plankton. Some of them even managed to crawl onto the soil.

In the same way that there is a natural tendency to view extinct creatures as failures, there is a natural inclination to view trilobites as failures, despite the reality that they were among the most successful creatures that have ever lived (Their reign lasted for 300 million years, which is twice as long as that of the dinosaurs, who were themselves one of the most successful species to ever exist. To this point, humans have only lasted for less than 1% as long as trilobites).[0]

In science, trilobites were nearly the only known forms of ancient complex life for the entirety of the 19th century. Hence, they were the subject of extensive research and collection efforts. Their unforeseen arrival was the primary factor in the intrigue surrounding them. Even to this day, it can be surprising to go to the right formation of rocks and make your way upward through the aeons finding no evident life at all, and then all of a sudden, a whole Profallotaspis or Elenellus (both extinct genera of trilobites) that is as big as a crab will pop everywhere. These weird creatures had limbs, gills, properly developed nerve systems, probing antennae, a small but complex brain, and the most bizarre eyes anybody had ever seen. They were made of calcite rods, which is the same material that is used to make limestone, and they were the oldest known sensory systems. In addition to this, the early trilobites didn't only consist of a single daring species but rather many, and they didn't only show up in one or two places but somewhat all over the world.

It's likely that the Cambrian explosion initiated a rapid rise in size rather than the abrupt introduction of new body forms. In addition to this, it occurred in a very short amount of time, indicating that it was an explosion. The theory proposes that just in the same way that mammals ended up waiting for hundreds of millions of years until the dinosaurs went extinct and then seemingly erupted in the vast array all over the entire globe, it is possible that arthropods and other triploblasts ended up waiting in semi microscopic secrecy for the dominant Ediacaran organisms to have their day.

Life is a peculiar phenomenon, especially when viewed from the standpoint of a human being, which, as obvious as it may be, is the vantage point from which we must examine it. Life couldn't wait to get started, but once it did, it appeared to be in very little rush to progress once it had already begun rolling.

In his book, *A Short History of Nearly Everything*,[0] Bill Bryson made a superb 24-hour analogy for understanding the timeline of life: if you believe that the roughly 4.5 billion years of Earth's history can be condensed into one day, then life on this planet begins exceptionally early, at 4 a.m., with the birth of the very first simple, single-celled organisms. However, life does not evolve any further for the following 16 hours! Not until almost 8:30 p.m., when the day has already been completed for five-sixths of its duration, does Earth have anything to show the cosmos other than a restless skin of bacteria. Then, after what seems like an eternity, the first sea flora makes its debut, quickly followed 20 minutes later by the first jellyfish and the mysterious Ediacaran fauna that Reginald Sprigg discovered for the first time in Australia.

However, Earth has to go through a lot to get to a point where life might exist. Nothing except a hellishly unstable climate plagued it for the overwhelming duration of its existence. Earth's surface was covered in lava and averaged 1,400 degrees Celsius when life first arose. By 9 p.m., or eight-tenths of the day had passed, oxygen levels had risen to the point where a protective layer of ozone could grow (the Ozone Layer), therefore absorbing part of the Sun's rays. Heavy UV light is deadly to plants and kills the plankton that feeds most of the ocean's population. In this way, the ozone layer made life's complex development possible.

At exactly 9:04 p.m., trilobites emerge from the depths of the Burgess Shale—a record of one of the earliest marine ecosystems—followed more or less immediately after by the elongated animals of the same formation. On the land, vegetation starts to appear just before 10 p.m., and the first land animals appear not long after, with just under two hours still left in the day. By 10:24 p.m., the Earth is covered in enormous carboniferous forests whose remnants provide us with all of our coal, and the first flying insects are seen. This is all owing to the very brief period of warm weather that occurred just 10 minutes earlier. Dinosaurs stomp onto the scene just a few minutes before 11 p.m., and maintain their dominance for around three-quarters of an hour. They disappeared at 11:39 p.m., marking the beginning of the Age of Mammals. Humans arise one minute and seventeen seconds before midnight. On this scale, the entirety of our documented history, which comprises the remaining chapters in this book, would equal little more than a few seconds, and your lifespan would be just an instant in comparison (If every letter in this book symbolises the 3.85-billion-year history of life, then one letter represents roughly 7,700 years!).

Continents tumble and collide with one other at a pace that appears volatile during this dramatically accelerated day. Glaciers advance and retreat, deep ocean form and disappear, and mountains expand and shrink throughout time. About three times each minute, somewhere on Earth, a bright flash of light marks the collision of a meteor of a Manson-sized meteor (2.4 kilometres in diameter). Incredibly, life is able to hold on in such a turbulent and harsh environment. One other approach to grasp the vastness of time is to envision your arms stretched out to their full length and the width of those arms, from left to right, as the planet's history. The entire intricate history of complex life just starts at your right hand, and human history will be wiped out with one swift swipe of a fingernail!

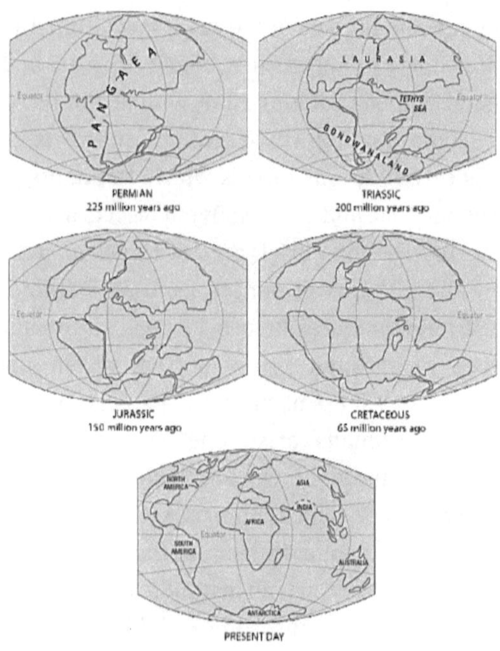

15) The breakup of Pangaea and motion of their continents to their present-day positions.

Thankfully, that day has not yet come, but it certainly might since there is one more crucial aspect of Earth's existence that is often overlooked: it eventually goes extinct (regularly, in fact). Species go to great lengths to develop and maintain themselves, yet in the end, they die and collapse with remarkable regularity (it is a natural principle). Their rate of extinction appears to increase in tandem with their complexity, which is perhaps why it's possible that much of life isn't really ambitious to develop. It's a big deal whenever life takes a risk, and the moment it emerged from the ocean and into the next phase of our story was no exception.

The land was a hostile place to be, particularly with its high temperatures and exposure to strong UV rays, not to mention the fact that you couldn't swim as you would in water. To adapt to life on land, animals had to drastically alter their anatomy. If you hold a fish by its tail and its fins, the fish will break in the centre because its spine isn't strong enough to keep it up. Adapting to life on land was not a simple matter for aquatic animals, as they were required to develop new load-bearing anatomies. Most importantly, any terrestrial species would need to figure out how to breathe in oxygen from the air instead of relying on gills and oxygen-rich water. These obstacles were not easy to handle. However, there was a compelling reason to get out of the water: it was becoming increasingly dangerous in the water. Because of the gradual merging of the continents into a single landmass called Pangaea, there is now dramatically less fertile marine territory than there used to be. As a result, the marine competition was quite intense, and the appearance of a new kind of predator, the shark, was also a significant development. This predator was both omnivorous and terrifying since it was tailored so flawlessly for an attack that it has altered very little through the aeons. So, in short, there has never been a better moment to look for habitats outside of the water!

The Biological Big Bang

Around 450 million years ago, plants began colonising land, and with them came the small insects and other animals they relied on to decompose and recycle dead organic materials. When we first look around the world, one of the most striking things is that the climate and other physical characteristics of different locations cannot explain the similarities and differences among their respective populations. Comparing wide swaths of territory in Australia, South Africa, and western South America between 25 degrees and 35 degrees south latitude, we shall discover areas highly comparable in all their conditions, yet it would be impossible to pick out three faunas and floras more radically distinct. Or, to take another example, we can see that the faunas and floras of southern South America (below latitude 35 degrees) are more closely related to each other than those of northern South America (above latitude 25 degrees), which inhabits a quite different environment.[2]

The second major insight that emerges from our comprehensive analysis is the direct connection between the variations in regional development and barriers to unrestricted migration. There are absolutely no two marine faunas more different from one another than the marine faunas of the eastern and western coasts of South and Central America, and yet these huge faunas are divided solely by the short, but the impenetrable land bridge of Panama (At its narrowest part, Panama is barely 80 kilometres wide!). After this barrier is crossed, we come upon the eastern islands of the Pacific, which are home to a whole different set of species than those found in the western islands of the Pacific. As a result, three unique marine faunas can be found in this region, all of which are found far to the north and south and on parallel lines not too far apart, with climates that are similar but not identical. Falling Pliocene temperatures provide more useful information on regional progress and obstacles to free movement. The common species found in both the New and Old Worlds must have been entirely cut off from each other as soon as they went south of the Polar Circle during the gradually declining temperature of the Pliocene epoch, the latter phases of global cooling.[2]

There is a third basic fact that is implicit in the two arguments above: the development of the same continent or sea is comparable even while the species themselves are different in different areas. Many of these characteristics can be passed down through many generations. For example, the skeletons of turtles, horses, humans, birds, and bats are quite similar. The corresponding bones are easily visible all throughout the body, not just in the limbs. From a purely pragmatic perspective, it seems inexplicable that the forelimbs of a turtle, horse, human, and bird or bat are all constructed from the same bones. Only if these species have a common ancestor and had only modest structural modifications to adapt to different settings can the similarities in skeletal structure be explained. As both points of view would have it, it's crystal evident that all species within the same genus share a common ancestor despite their geographically dispersed distribution. For species that have persisted through whole geological epochs with little modification, almost any degree of migration is possible due to the immense geographical and climatic upheavals that have supervened since prehistoric times.

Any biologist who has travelled, say, from the Arctic to the Antarctic would have been startled by the rapid succession of radically different but clearly related groups of organisms as they replaced the ones that came before them. The naturalist hears from related but unique bird species, makes nearly identical observations, and observes nests with nearly identical but slightly different egg colours when travelling from north to south. The islands off the American coast all have slightly different geological structures, but their unique bird species are essentially all distinct. These findings indicate a profound organic connection, one that holds true through vast distances and aeons, over the same stretches of land and ocean, and regardless of the physical conditions of these regions.

However, among the numerous difficulties posed by plant distribution patterns is the fact that many animal species are confined to single continents, while plant species have disjunct distributions as they are found in different parts of the world. However, Darwin provides some ideas for why this may be possible.

In one of his experiments, out of 87 species, Darwin observed that 64 germinated after 28 days in water, and a handful survived after 137 days. For practical reasons, Darwin only tested little seeds without the casing or fruit. All of them sank after a few days, making it impossible to float them over great expanses of the ocean, regardless of whether or not they were damaged by the saltwater. So, the saltwater was where Darwin found the dried stems and branches of 94 plants with ripe fruit. Most of them rapidly sunk, but others floated briefly when young but for substantially longer once dried. For example, ripe hazel-nuts sank the moment they were picked, but when dried, they floated for 90 days. Therefore, it would be reasonable to anticipate that around 10% of a flora's seeds—after being dried—could well be transported over a stretch of ocean 1,500 kilometres wide and would subsequently germinate.[2]

Birds are another method through which a flora's plants might "rapidly" spread. Many terrestrial and aquatic plant seeds may be found in the diets of freshwater fish (as birds regularly consume fish, this suggests that the seeds may be spread from place to place). There's no way that birds won't be very efficient seed distributors. After a bird devours a large supply of food, it is strongly concluded that all the grains do not pass into the gizzard for 12 or perhaps even 18 hours. A bird flying in this territory may travel up to 800 kilometres away.[2]

It would be a remarkable reality if many plants did not get extensively disseminated in one of these ways, given that the various aforementioned modes of transport have been in motion for centuries, millennia, and millions of years.

However, coming back to the diffusion of life, it took longer for larger animals to leave the sea and expand on land, but they began doing so some 400 million years ago. Many depictions of the early land inhabitants have us picturing an amphibious creature or some sort of fish with an adventurous spirit, like the present mudskipper, which may go from pond to pond during droughts. Indeed, modern woodlice, sometimes known as pillbugs or sow bugs, are presumably much more like the earliest obvious mobile creatures of terrestrial land.

To those who figured out how to breathe in oxygen, life was wonderful. When terrestrial life began flourishing in the Devonian and Carboniferous periods (about 390 million years ago), oxygen levels reached as high as 35% (as opposed to nearer 20% now). Because of this, creatures could reach astonishingly great sizes in a short amount of time. Extremely high oxygen levels were obviously beneficial, as they led to enormous size. Nearly one metre in length, a fossilized

16) The amphibian-like *Pederpes*, the most primitive tetrapod found in the Mississippian, and known from Scotland.

millipede-like creature's footprint left on a rock in Scotland from approximately 350 million years ago seems to be the earliest evidence of an animal living on land discovered to date. Some millipedes might even quadruple that size before the era ended. It's not too unexpected that insects of that time learnt to fly in order to avoid being eaten by such predators, given the prevalence of such animals at the time. Some took to this new method of transportation with such ease that they haven't modified their methods since they initially started developing and using them.[0]

Yet, a lot is unknown about the first terrestrial vertebrates, from which humans might well have derived. In the 1930s and the 1940s, Jarvik travelled to Greenland with a group of other Scandinavian scientists in search of fish fossils. They were on the lookout for lobe-finned fish, the presumed ancestors of tetrapods like humans and other walking animals. The vast majority of animal species are tetrapods, and all existing tetrapods have a single defining characteristic: they all have four limbs that end in no more than five toes or fingers. All tetrapods—dinosaurs, whales, birds, and even fish—share a common ancestor with all other vertebrates. The Devonian period, which occurred around 400 million

17) *Tyrannosaurus rex* (rex meaning "king" in Latin) is one of the most well-represented theropod species. It is regarded as the most well-known dinosaur, partly because it was the first extremely large carnivorous dinosaur to be discovered (long time it was believed to have been the largest ever to have lived) and that it was likely the most voracious meat eater in history.

years ago, was thought to have a trace to this predecessor. None of today's land animals existed before it. Following that period, a great deal had occurred. Yet, the team was fortunate enough to stumble upon an *Ichthyostega* (an extinct genus of limbed tetrapodomorphs), a marine animal measuring in at just over one metre in length (In the three early tetrapods that have been discovered so far, not a single one had five toes or fingers. In a nutshell, we have no concrete idea of our origins).

Each of these main classifications eventually gave rise to subcategories, some of which were successful while others were not. The turtles developed from anapsids and, for a short, maybe unlikely moment, looked set to reign as the world's most powerful and ruthless animal, before a change in evolution led them to pick endurance over dominance.

During the Permian epoch, only one of the four subgroups of synapsids survived. Lucky for us, this branch of the evolutionary tree led to the therapsids, a group of protomammals.

Unfortunately for the therapsids, their relatives, the diapsids, were also fast evolving. In their case, they became dinosaurs (among many other animals), which ultimately proved to be too much for the therapsids to handle. The therapsids have mostly disappeared from historical records because they were unable to compete successfully with the more aggressive new species. However, a very tiny fraction of them evolved into little, fluffy, digging animals that waited a tremendously long time as compact tiny mammals. These species had a lot of hair, and the largest of them was not much bigger than a house cat (most of them were hardly bigger than mice!). However, ironically, this would prove to be their saviour in the long run, but they would have to wait roughly 150 million years for the Age of Dinosaurs to unexpectedly come to an end so that the Age of Mammals could begin.

In addition to the innumerable smaller ones that took place between and after them, each of these tremendous transformations was reliant on one of the most ironically important forces of evolutionary development: extinction. It is a peculiar reality that on Earth, the extinction of species is, in the strictest sense, a way of life. This is an interesting truth. Nobody really knows how many different types of species have existed since the beginning of life. The total has now been estimated to be anywhere from 30 billion to 4,000 billion, with the latter being the more typical estimate. Regardless of the precise number, 99.99% of all species that have ever existed have become extinct since the beginning of time. The average expected lifespan of a species is only approximately three to four million years, which is roughly the same amount of time that has passed since humans first appeared on the planet (However, the complexity of a species greatly influences its longevity, with the more complex, the quicker its demise). Because the opposite of extinction is stagnation, extinction is almost certainly devastating news for the beings who are going to be affected by it, but it generally is a good development for a world that is constantly evolving.

The Ordovician, Devonian, Permian, Triassic, and Cretaceous mass extinctions (the Big Five) are the most well-known, although there have been several more (smaller) extinction events on Earth throughout its history. As much as 85% of all species disappeared throughout the Ordovician (440 million years ago) and Devonian (365 million years ago). About 75-85% of all species disappeared during the Triassic (210 million years ago) and the Cretaceous (65 million years ago) extinction events. The actual killer, though, was the Permian extinction, which occurred 245 million years ago and opened the door to the dinosaurs' uninterrupted reign. At least 95% of all animal species proven by fossils disappeared in the Permian and have not been seen since. This was the only major extinction event in insect history, and it wiped out nearly one-third of insect species. According to Bryson, "this is the closest life has ever gone to complete extinction".[0]

The Hemphillian, Frasnian, Famennian, and Rancholabrean events, and a handful more are examples of relatively small, less well-known extinction events that occurred in between the larger ones and were not as catastrophic to total species levels but frequently badly damaged certain groups. About five million years ago, the

Hemphillian catastrophe almost extinguished all grass-eating creatures. There is just one surviving horse species, and its fossil records are so sparse that it may have been on the verge of extinction for a while (picture human civilisation developing in the absence of horses and other grazing animals).

For both large and small extinction events, our lack of knowledge about their causes is baffling. After eliminating the most far-fetched suggestions, there remain more explanations than extinction events. Many different explanations or primary have been outlined, including but not limited to: global warming, global cooling, adjusting ocean levels, oxygen depletion of the oceans (a condition known as anoxia), epidemics, gigantic leakages of methane gas from the seafloor, meteor and comet impacts, rapidly escalating storms of the known as "hyper canes", huge volcanic upwellings, and disastrous solar flares.

The last idea, in particular, is noteworthy. As the Sun is a powerful machine, so too are its outbursts. Yet, no one really understands just how massive solar outbursts may grow because we have only started studying them ever since the beginning of the Space Age (1957). We on Earth wouldn't really feel a normal solar flare, yet the power it releases is comparable to a billion hydrogen bombs. It is believed that a really powerful blast, perhaps many hundreds of times the ordinary flare, may surpass our ethereal defences. The magnetosphere and atmosphere between them generally sweep them back into space or guide them gently towards the poles (where they form the Earth's beautiful and charming auroras). Even while the light show would be spectacular, it would definitely kill the majority of the individuals who looked at it. What's even scarier is the fact that such as storm would leave no mark in history.

One scholar has summarised the state of the field as "loads of speculation and minimal proof." At least three of the most significant extinction events, the Ordovician, Devonian, and the Permian, seem to be linked to cooling, but apart from that, very little is accepted, including the speed with which a certain occurrence unfolded. For example, scientists disagree if the late Devonian extinction, which was followed by the terrestrialisation of vertebrates, occurred over millions of years, millennia, decennia, years or just in one single day.

The Age of Mammals

It is extremely challenging to eradicate all life forms from the face of the Earth, which is one of the main reasons why extinction theories are so difficult to come up with. But then why was the Cretaceous–Tertiary extinction (the extermination of the dinosaurs), also known as the KT event, during the Cenozoic Era so exceptionally catastrophic as compared to the myriad of other impacts our planet has experienced?

In the first place, the KT event was monstrously large. The Chicxulub impact crater is located beneath the Yucatán Peninsula in Mexico. Its centre is located offshore near Chicxulub, after which it is named. It was formed approximately 66 million years ago when a 10-kilometre-wide asteroid collided with Earth. It is estimated that the crater had a diameter of 180 kilometres and a depth of 20 kilometres (debris and sediment have since filled the crater).

18) Panorama of Meteor Crater in the Winslow region of Arizona. This 1,186-metre-diameter crater was created approximately 50,000 years ago during the Pleistocene epoch by a 50-metre-wide nickel-iron meteorite. With an impact energy of approximately ten megatons of TNT, this impact was roughly ten million times smaller (in terms of energy) than the KT impact, but it still produced a massive crater.

The impactor's projected speed of 20 kilometres per second was enough to produce an impact energy equivalent to 72–100 teratonnes of TNT (one teraton equals a million megatons). There are many potential repercussions, but one of them is that the collision led the Earth to be plunged into complete darkness for at least two years. This is only one of the conceivable outcomes. Research published in the journal *Geophysical Research Letters* concluded that the average temperature in the tropics dropped from 27 degrees Celsius to 5 degrees Celsius during the Paleogene era, a subdivision of the Cenozoic epoch (65.5–23 million years ago).[0]

Although it's hard to fathom such a massive explosion, American geoscientist James Lawrence Powell has recently noted that even if you detonated a Hiroshima-sized nuke for every person on the planet today, you'd still be roughly a billion bombs short of the magnitude of the KT catastrophe. However, even with such colossal numbers, that might not have been quite enough to wipe out 70% of all life on the planet.

What caused the extinction of 70% of all species on Earth during that time? That being said, the survival of the other 30% is an even more compelling mystery. How is it possible that all dinosaurs died out, whereas other reptiles, such as snakes and crocodiles, were unaffected? No known frog, toad, salamander, or other amphibian species in North America went extinct. This was also true of the oceans. Despite the extinction of the ammonites (an extinct subgenus of the squid), their aquatic relatives, the nautiloids, survived. Whilst many plankton species, namely 92% of foraminiferans, were pretty much eradicated, others, like diatoms, which were similarly built and lived nearby, were relatively undamaged.

Nearly 90% of land-based organisms perished throughout the KT event, but only 10% of freshwater species perished. Water not only served as a shield against the raging fire, but also, apparently, as a source of increased nutrition during the subsequent famine. All the terrestrial creatures who made it through the struggle had a common survival strategy: they would flee to the safety of freshwater, little ponds and rivers, or the deep underground when danger loomed. Scavenging animals might have benefited from this as well. Lizards were and still are resistant to the germs found in decomposing meat. In fact, they are typically attracted to it, which is evident since, after the impact, there were a large number of decomposing carcasses lying around for a considerable amount of time.

The idea that only tiny creatures survived the KT event is untrue. Crocodiles were some of the survivors, and they were enormous compared to modern-day standards. However, it is generally true that almost all survivors were rather little and sly. Our mammalian ancestors thrived in this harsh environment because they were tiny, warm-blooded, nocturnal, dietary chameleons, and naturally wary. We likely would have perished if our development had progressed any further. Instead, mammals awoke on a planet that was just as conducive to their survival as any other living creature. Rapid mammalian evolution just after dinosaurs died out covered the void left by these ancient mythical creatures.

Since we've gotten used to thinking of ourselves as life's inevitable dominating species, it's difficult to comprehend that we've been only here due to cosmic blasts at the right time and a few other lucky events. For almost four billion years, our predecessors have successfully slipped through a chain of nature's evolutionary selection pressures every time they had to. This is the fundamental thing we have in common with all living beings. As American palaeontologist Stephen Jay Gould famously put it, "Humans exist today because our unique lineage never fractured—never once at any of the billion points that could have eliminated us from history."

So we can conclude by now that the Cretaceous–Tertiary extinction in the Cenozoic Era truly had two-sided consequences. (Speaking of the Cenozoic Era, this era was the third of the Earth's major eras, and it was during this era that the continents took on their current shapes and placements). The Cenozoic epoch, named after the Greek word for "recent life," began approximately several pages ago, or about 65 million years ago, and is split into the Paleogene (65.5–23 million years ago), Neogene (23–2.6 million years ago), and Quaternary eras (2.6 million years ago to the present).[4]

The Cenozoic Era saw the development of many of the world's biggest mountain ranges. By up-warping the streambed, the slopes became steeper, giving the rivers more erosive force. Therefore, throughout the late Cenozoic, deep river valleys and canyons, like the Grand Canyon of the Colorado River in northern Arizona, were carved into vast upwarps of sedimentary rock. A change in air circulation and weather patterns driven by the new mountain ranges resulted in a drier and colder environment. With this change, there was an increase in the size and thickness of the Arctic ice cap. Water remote from the seas got trapped in snow and ice on the high mountains. As the ocean receded, land connections between Africa and Eurasia, as well as Eurasia and North America, became possible.

The slow evolution of current forms of the world's flora and fauna began during this era as well, mostly due to the fact that the linking of the continents made new land available to organisms that had evolved in isolation. From Africa, apes and elephants spread over Eurasia. Africa became the new home for rabbits, pigs, sabre-toothed cats, and rhinos. After crossing the Bering Strait (the frozen land bridge that formerly connected East Russia and the American state of Alaska), elephants and rhinos eventually arrived in North America (Interestingly, horses headed in the opposite direction: from Alaska to East Russia). As raccoons made their way north, ground sloths were making the reverse journey from South America to North America. It's possible that rodents made their way from Southeast Asia to Australia by way of the Pacific Islands.[4]

Grasslands, better adapted to the new colder and drier conditions, gradually replaced many of the vast forests that formerly covered the continents from coast to coast and from Pole to Pole. Animals that consume plants had to change their diets to keep themselves alive. As a result of selective pressure, horses developed thicker enamel to better protect their teeth. In the same vein, ruminants like bison, camels, sheep, and giraffes ate grass because their stomachs are divided into sections that are more suited to breaking it down. The grazers' newfound agility and herd behaviour in the field was a great tribute to their adaptability. Predators had to change as well, and meanwhile on land, Asian and African apes diverged from proto-primates and then, several million years later, hominins split from their closest African ape ancestors, the chimpanzees.

The first primate-like mammals ever discovered are called proto-primates. They resembled a rodent that had been created by crossing a squirrel and a mouse in terms of size and shape. The nature of these proto-primates will remain a mystery until further fossil information is unearthed, so long as the enigma persists. *Plesiadapis* is the first primate-like animal for which there is a fossil record that can be considered substantially comprehensive (although some researchers do not agree that *Plesiadapis* was a proto-primate). The earliest evidence of this primate found in fossils dates back 55 million years. During the Cenozoic, they were present in Europe and North America, but by the end of the Paleogene, they were no longer present anywhere on Earth.[5]

During the Paleogene, the first real monkeys appeared on each of the four continents. These ancient primates were similar to present prosimians like lemurs and other prosimians. The brains and eyes of these early primates

19) Life restoration of *Plesiadapis*. The first discovery of *Plesiadapis* was made by François Louis Paul Gervaise in 1877, who first discovered *Plesiadapis tricuspidens* in France.

gradually became larger as they developed, but their muzzles became shorter. The bulk of the early prosimians became extinct before the end of the Paleogene, most likely as a consequence of changes in temperature or as a result of competition with the first monkeys.

Prosimians, which lived during the Oligocene Period, are considered to be the ancestors of anthropoid monkeys, which are found today. There is evidence that by 40 million years ago, monkeys lived in both Africa and Asia and South America. By the time the platyrrhines appeared on the scene, the continents of South America and Africa had already begun to migrate apart. It is generally accepted that monkeys evolved in the Old World and subsequently moved to the New World by traversing land bridges between the two regions. Due to the fact that New World monkeys and Old World monkeys were unable to breed with one another for millions of years, each subspecies of monkeys in its own regions underwent separate patterns of evolutionary adaptation.

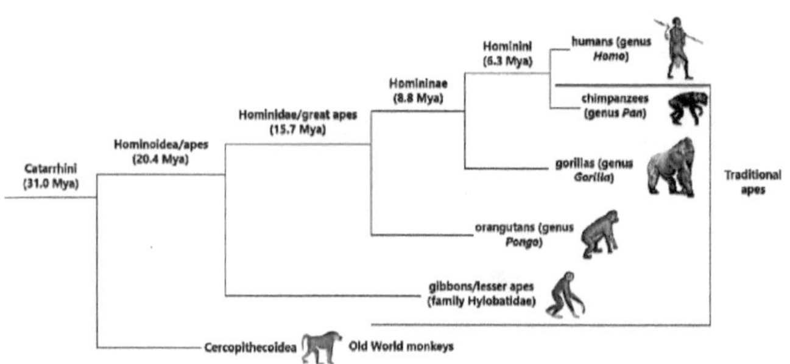

20) Taxonomic classification and phylogeny of the superfamily Hominoidea.

Although all monkeys in the New World make their homes in the trees, certain species of Old World monkeys are capable of both living on the ground and doing impressive tree-to-ground acrobatics.[5]

Approximately 31 million years ago, in Africa, apes and catarrhines began to split from one another, marking the middle of the Cenozoic epoch. The average size of an ape is far larger than that of a monkey, and in addition, they do not have a tail. All apes have the capability of traversing among the branches of trees, despite the fact that many ape species spend the large bulk of their time on the ground. Because of the larger size of their brains in comparison to the rest of their bodies, apes are superior to monkeys in terms of intelligence.

There are two main groups of apes living in the world today: Gibbons and siamangs are both members of the family Hylobatidae, which is comprised of the more little apes. The great apes include the genera *Pan* (chimpanzees and bonobos), *Gorilla* (gorillas), *Pongo* (orangutans), and *Homo* (humans). The family Hominidae, which is classified under the Primates division, is home to the great apes, sometimes referred to as hominoids. Both the fossil record and a comparison of the DNA of humans and chimpanzees lend credence to the theory that humans and apes diverged from a common ancestor of the hominoid species around six million years ago.[5]

Understanding that we, as humans, are not at the pinnacle of anything is one of the most difficult ideas for our species to grasp. We did not have to come here, yet our presence was still unavoidable. Because of our inflated sense of self-importance, we tend to see evolution as if it were a computer programme designed to create us.

As we will see in the book's remaining chapters, the next period, also known as the Quaternary era (2.6 million years ago to the present), is a lengthy and contentious discussion about what happened next in the history of human development. But before we go any further, it's important to realise that after about six million years of evolutionary wobbling, from faraway, befuddled early hominins (*Australopithecus*) to you, we ended up with a species that is still 98.4% genetically indistinguishable from the present chimpanzee. There is a greater similarity between you and the hairy creatures your distant ancestors left behind many million years ago when they started to walk on two legs than between a zebra and a horse or a dolphin and a porpoise.

This part and its chapters of the book began with a discussion of three main ideas: life has the desire to exist, life does not always have a strong desire to evolve, and life can go extinct at any time. To this, we may add a fourth, which is that life goes on regardless. And frequently, as we shall see, life continues in ways that are utterly astounding, to say the least.

PART II

THE COGNITIVE REVOLUTION

0) The paintings, dated to between 15,000 and 17,000 years ago and titled "Aurochs, Horses, and Deer," are among of the greatest examples of Upper Paleolithic art, and they depict primarily animal subjects.

The Birth and Rise of *Homo sapiens*

4. THE ADVENT OF SAPIENS

Never before have we lived in a world so carefully organised and technologically advanced. Many humans have access to shelter, food, safety, and comfort and no longer have to worry about their survival, which is now taken for granted. Yet, as we shall see in the next parts and chapters, a thorough examination of the glorious history of mankind reveals how fast our knowledge, technology, power and ego have risen. For almost 99.99% of human history, life was completely different. In reality, however, there is no "one" human history. We have a tendency to believe that we, *Homo sapiens*, are the only human species that has ever existed, but this is not true and conveys an ignorant disdain for our brilliant ancestors and distant relatives.

The Human Family Tree

As science advances, we are discovering evidence that the human species is considerably older than previously believed. In the past 25 to 30 years, experts have stated that anatomically modern humans did not exist until 50,000 years ago. Then, just recently in Morocco, scientists discovered 300,000-year-old evidence of anatomically modern humans with the same brains, capacities, and abilities as us. According to Professor Yuval Noah Harari, author of the famous intriguing book *Sapiens: A Brief History of Humankind*,[0] humans descended from an older genus of apes named

Australopithecus in East Africa approximately 2.5 million years ago (also known as the Southern Ape). The brains of most Australopithecus species were around 35% the size of a modern human brain. Although this is greater than the average chimp endocranial volume, the earliest australopiths (*Australopithecus anamensis*), a hominin species that lived between 4.2 and 3.8 million years ago, appear to have been within the chimp range, whereas some later australopith specimens have a larger endocranial volume than some early *Homo* fossils. While none of the groupings survived, Australopithecus did produce living offspring, as the genus *Homo* arose from an Australopithecus species some 2.5 million years ago. These australopith species have been postulated as the ancestor or sister of the *Homo* lineage due to physical similarities, although there is no agreement on which gave rise to *Homo* (*Homo* in Latin signifies "human being" or "man" in the broad sense of "human being" or "mankind").

Approximately 2.1 million years ago, *Homo habilis* arose as one of the earliest living *Homo* descendants of Australopithecus. Even before 2010, there were suggestions that *Homo habilis* should not be placed in the genus *Homo*, but rather in *Australopithecus*. The main reason for including *Homo habilis* in *Homo*, the undisputed use of tools, has been superseded by the discovery of the use of *Australopithecus* tools at least a million years before *Homo habilis*. Several specimens with insecure species identification were assigned to *Homo habilis*, leading to arguments for splitting into *Homo rudolfensis* and *Homo gautengensis*, of which only the former has received broad support. In addition, *Homo habilis* was long thought to be the ancestor of the more graceful *Homo ergaster* (*Homo erectus*).

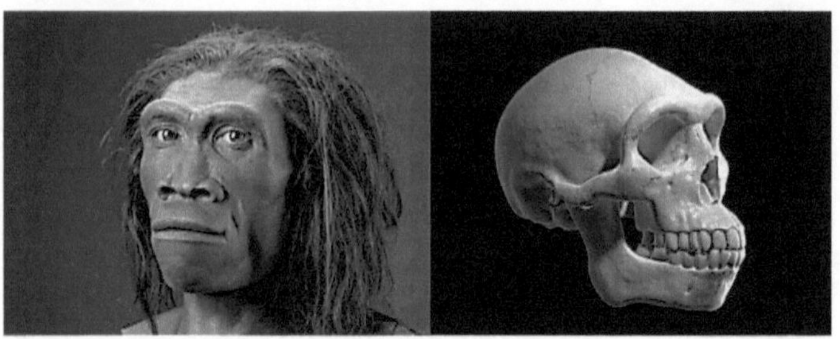

1) Reconstruction of *Homo erectus*. From around 1.9 million years ago, the species rapidly migrated over the African tropics, Europe, South Asia, and Southeast Asia before becoming extinct 200,000 years ago.

However, some archaic humans left their homeland approximately two million years ago to journey through and inhabit broad areas of Africa and Eurasia (these expansions are referred to as "Out of Africa I"). Because life in the frigid woods of northern Europe required different features than survival in the sweltering hot jungles of South-East Asia, human populations developed in diverse directions, eventually resulting in unique species, to which scientists have assigned pretentious Latin names.[0]

Homo erectus (nicknamed "Upright Man") inhabited eastern Asia and thrived there for approximately two million years, making it the most resilient human species yet. In comparison, *Homo sapiens* has only been around for roughly 200,000 years. Several human species, including *Homo heidelbergensis* and *Homo antecessor*, appear to have evolved from *Homo erectus*. The former is usually thought to be the ancestor of Neanderthals, Denisovans, and modern humans. But returning to *Homo erectus*, this endurance milestone is

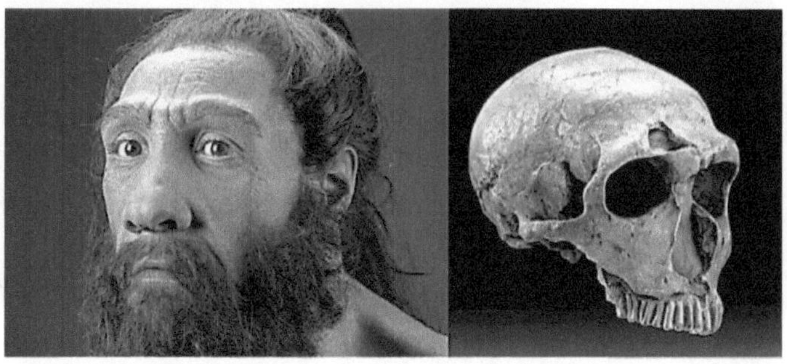

2) Reconstruction of *Homo neanderthalensis*. They emerged during the Pleistocene Epoch at least 200,000 years ago and were displaced or assimilated by early modern human species (*Homo sapiens*) between 35,000 and 24,000 years ago.

unlikely to be broken, even by *Homo sapiens*. *Homo erectus* walked with a more contemporary stride and was the first human species to have a flat face, large nose, and probably scant body hair.

A subspecies of *Homo erectus*, which inhabited the Zhoukoudian Cave of northern China during the Middle Pleistocene, was retrieved from the ground in 1921. This subspecies is also known as *Homo erectus pekinensis* (also nicknamed "Peking Man"). *Homo erectus*, a woodland and savannah biomes specialist, likely went extinct with the takeover of tropical rainforests. Peking Man's final stay at Zhoukoudian may have taken place sometime between 400,000–230,000 years ago.

Humans in Europe and Western Asia eventually evolved into *Homo neanderthalensis*; they were substantially more robust, muscular, and larger than *Homo sapiens* and were far better adapted to the frigid temperature of Ice Age western Eurasia. Neanderthals had substantially thicker and stronger bones than *Homo sapiens*. These bigger bones included thicker metacarpals and a more robust nature that matched their harsh lifestyle. Despite having substantially stronger and thicker bones, Neanderthals were subjected to a high incidence of traumatic injury, with an estimated 85% of specimens exhibiting signs of healed major trauma, with over 45% seriously harmed and nearly 20% injured before reaching adulthood.[0]

One extreme example is Shanidar 1, who shows signs of an amputation of the right arm, likely due to a nonunion after breaking a bone. Additionally, scientists examined his skeletal remains and found evidence that Shanidar 1 experienced a crushing blow to his head at a young age. The impact damaged his left eye (possibly blinding him) and the brain area controlling the right side of his body, resulting in a withered right arm, possible paralysis, and a crippled right leg.[1]

Homo soloensis (nicknamed "Man from the Solo Valley"), a tropic-adapted subspecies of *Homo erectus*, lived on the Indonesian island of Java. The Solo Man skull is oval-shaped in top view, with heavy brows, inflated cheekbones, and a prominent bar of bone wrapping around the back. On another Indonesian island (the small island of Flores), archaic humans underwent a process of dwarfing. Flores was initially visited by humans when sea levels were abnormally low. As a result, the island was easily accessible from the mainland. However, when the sea level rose, some humans were trapped on the poor-resourced island. The larger individuals perished first because they required more food and resources to survive. Over generations, evolutionary pressures reformed

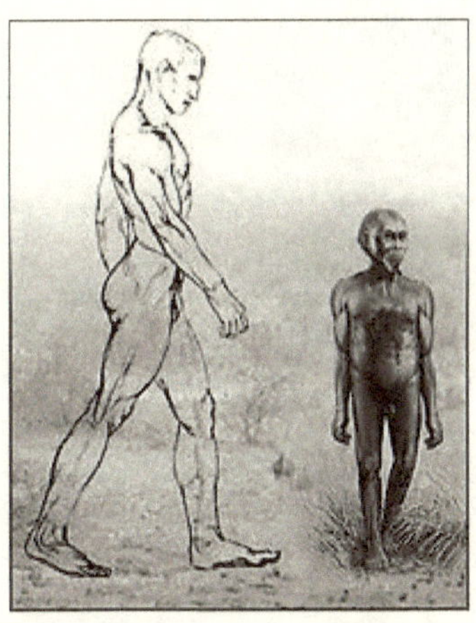

3) An illustration of a comparison between an adult male *Homo sapiens* (left) and an adult male *Homo floresiensis* (right).

the island's population to just dwarfs. This one-of-a-kind species, also known as *Homo floresiensis* ("Flores Man," also called "Hobbit"), stood barely one metre tall and weighed little more than 25 kilos, which for reference, is around the same height and weight as an average sea otter (*Enhydra lutris*), or the same weight as an 8-years-old modern *Homo sapiens*.

In 2010 another lost human *Homo* sibling was rescued from oblivion when scientists identified a Denisovan individual based on mitochondrial DNA from an earlier discovered fossilised finger bone from the Denisova Cave in Siberia. Because of the scarcity of physical remains, nothing is known about the Denisovans' unique anatomical traits. The finger, however, belonged to a previously undiscovered human species known as *Homo denisova*, according to DNA analyses.[0]

4) A reconstruction of the Hominini family tree, showing some of our most distant and oldest ancestors (*Australopithecus*, *Paranthropus*, and *Kenyanthropus*) and to some of our closest relatives (*Homo*).

We have covered almost the entire *Homo* family tree thus far, and it's almost unreal to consider and assume that there may still be a possibility that there are even more unknown lost fossilised relatives of ours lying underground, waiting to be discovered.

However, as these humans evolved in Europe and Asia, East African evolution continued. Many new species were formed in the birthplace of mankind, including *Homo redolfensis*, *Homo ergaster*, and finally our own species, *Homo sapiens* (labelled as "Wise Man"). Because of evolutionary pressures, some of the species' members differed greatly. Some were huge, while others were small. Some were fearful, while others were modest plant-gatherers. Others lived on a small little island, while others travelled and spread over continents. However, they all belonged to the species *Homo*, which meant they were all humans.

It's a typical misconception to think of these species as having a direct line of ancestry, with Ergaster giving birth to Erectus, Erectus giving birth to the Neanderthals, and the Neanderthals giving birth to Sapiens. This linear model conveys the incorrect impression that there was only one type of human on Earth at any one time, and that past species were simply older versions of ourselves. The fact is that the Earth was home to dozens of human species at the same time, from roughly two million years ago to around 10,000 years ago.[0]

By the time *Homo sapiens* arrived on the scene some 300,000 years ago, we were the ninth *Homo* species, joining *habilis*, *erectus*, *rudolfensis*, *heidelbergensis*, *floresiensis*, *neanderthalensis*, *naledi*, and *luzonensis*. Our current exclusivity, rather than the multi-species past, is unusual and possibly onerous. And as we will discover shortly, we Sapiens have strong reasons to ignore our siblings' memories.

The Cost of Thinking

Regardless of their variations, all human species share various distinguishing qualities and characteristics. The first and possibly most noteworthy feature is that humans have relatively large brains in comparison to other animals. While animals weighing 60 kilos have an average brain capacity of 200 cubic centimetres, early humans from around 2.5 million years ago had brains that were approximately 600 cubic centimetres in size (modern *Homo sapiens* have a brain averaging 1,200–1,400 cubic centimetres). The ratio of brain mass to total body mass tends to decrease as an animal's size increases. Large whales have extremely small brains relative to their body mass, but small rodents such as mice have relatively large brains, resulting in a brain-to-body mass ratio comparable to that of humans. At 1:7, ants have the highest brain-to-body mass ratio of any animal (*Homo sapiens* have a 1:40 ratio). Nevertheless, so only a few species surpass the human brain-to-body mass ratio, which raises an extremely interesting question: why are relatively large brains such an uncommon occurrence in the animal kingdom?

We are so proud of our high-level intelligence that we assume that when it comes to cerebral power, more must be better. However, having a big brain is, first and foremost, a costly evolutionary investment. The first is that a large brain is, first and foremost, a massive drain on the body and is impractical to carry around, especially when housed inside a giant hard skull.[0] For *Homo sapiens*, the brain accounts for only about 2–3% of the total body weight but consumes a quarter of the body's energy when the body is at rest. That implies that a human requires around 320 calories on an average day just to "think" (in fact, an average 5-6-year-brain old's may consume up to 60% of the body's energy!). In comparison, other apes' brains use "just" 8% of rest-time energy. As a result, our brain's principal function (processing and transferring information via electrical impulses) is extremely energy-intensive.

The precise percentages are hard to ascertain, but we have solid approximations of where that energy is flowing, which varies by brain region. The brain has nearly 100 billion nerve cells, and an unimaginable number of connections between them take to make up our neural networks. It's been estimated that there are a thousand million connections in a piece of brain the size of a grain of sand! In general, a quarter of the energy in the cerebral cortex of *Homo sapiens* goes to sustaining the neurons and glial cells themselves (the processes all cells go through to remain alive). The remaining 75% is dedicated to signalling (the transmission and processing of electrical impulses throughout the brain's circuits).[3]

Unlike muscles that can retain surplus carbs, the brain needs continual oxygen and energy supply to function correctly (thus, the brain doesn't have a reserve of energy to store away when needed). When the blood supply to the brain is cut off or interrupted, as in a stroke or head injury, neurons begin to die down rapidly. This may appear to be a flaw, yet it is necessary for the brain to function efficiently. If the brain has backup power cells, such cells will take up space between neurons, which would lengthen the distances that electrical impulses would have to travel, necessitating additional energy. Early species may have possessed neural systems with this kind of fail-safes, but humans lost this backup power for efficiency through millions of years of evolution. It makes us more vulnerable to harm, but it also allows us to benefit from the brain's sophisticated circuitry.[3]

The synapses absorb the majority of that energy (the tiny gaps between brain cells where signals are sent and received). There, the cells are constantly pushing ions into the gap between them (exchanging potassium and sodium to create electrical charges). This pumping activity is essential to the working of brain circuits, but it consumes a lot of energy.[3] Today, our huge brains pay off spectacularly, thanks to our ability to create complex and hazardous technologies.

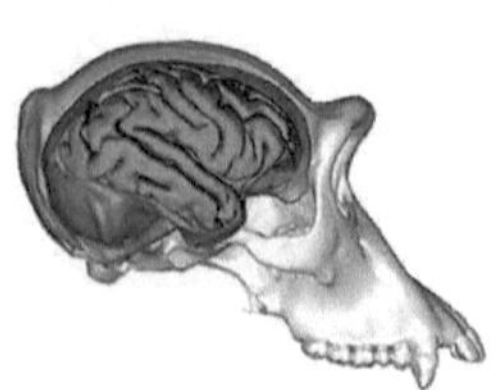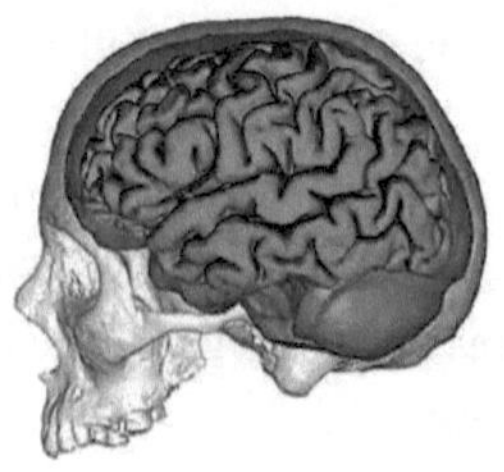

5) Three-dimensional reconstructions of the endocranial brains of a chimpanzee (left) and *Homo sapiens* (right) skull. Whereas *Homo sapiens* have an average brain size of 1,200–1,400 cm^3, chimpanzees have an average brain size of 400 cm^3.

However, while human brain networks grew and grew for more than two million years, humans still had little to show for it other than some flint blades and sharp sticks.[0]

So despite our "extraordinary" big brains, it took humans several million years to get to a prosperous and organised civilisation. Having giant brains does not mean you get to win automatically. Perhaps our profound intelligence and big brains are simply not so great, and we are lucky it eventually worked out for us. So what then drove forward the evolution of our massive human brain during those two million years? Frankly, we don't know exactly, but we have several possible theories.

The human brain, like a vehicle engine, performs best when kept at a constant cool temperature. If an animal's brain isn't working properly, then the body it controls won't work properly either. Therefore, in the fiercely competitive natural world, having a brain that overheats is a guaranteed way to extinction. Natural selection may have favoured those with a larger network of emissary veins inside the brain, allowing for greater dissipation of heat. This evolutionary development may have paved the way for bigger brains as a result of evolution.

However, mastering fire seems to be the most plausible explanation for the acceleration of evolution toward larger skulls and brains. Some human species may have used fire for the first time as early as 800,000 years ago. By 300,000 years ago, Erectus, Neanderthals, and Sapiens relatives were regularly utilising fire. Cooking with fire makes food significantly more nourishing and nutritious. Furthermore, it converts crops that humans cannot digest in their natural forms (such as wheat, rice, and potatoes) into dietary staples. Cooking also kills bacteria and parasites present in foods. Cooked fruits, nuts, insects, and carrion are also much easier for humans to chew and digest. Whereas primates chew raw food for extended periods of time, humans just need an hour to devour prepared food.[0] Obtaining drastically more nutritious food was a significant milestone, which many scientists believe developed the brain significantly. The evolutionary pressure on our brains to increase in size presumably resulted from "expertise" in conjunction with this seismic shift. More knowledge and experience improved chances of survival on the savannah, which in turn necessitated a larger brain.

Aside from the development of the brain, fire also produced a dependable source of light that kept humans warm, which made winters less gruesome. Fire also was used for hunting and protected us against enemies and predators (Can you imagine how different the journey of *Homo sapiens* would have been if they hadn't mastered fire?)

Because practically all animals rely on their bodies for power (height, muscles, size of their teeth), fire formed the first big divide between humans and other animals.[0] Of course, crocodiles, for example, have powerful jaws with many conical teeth, thick plated skin and can walk on land; yet, their maximum walking speed is strictly proportional to their short legs with clawed webbed toes. However, when humans domesticated fire, they gained complete control of an obedient and potentially limitless force they could ignite whenever needed. Thus, without a doubt, fire truly gave us unlimited power.[2]

Humans may have even started deliberately setting fire to their neighbourhoods not long after. A well-managed fire may transform impenetrable barren vegetation into great game plains. Furthermore, humans could stroll through the ruins and collect burned animals and nuts when the fire had been quenched. Cooking helped humans to start eating a wider variety of foods, spend less time eating and make do with smaller teeth and shorter intestines. Aside from brain development, experts believe there is a clear relationship between the invention of cooking and the shortening of the human digestive system.

The Distinctiveness of Sapiens

However, going back to the significant distinguishing features among the human species, in addition to our relatively large cerebrum, the second distinctive feature of the human species is that we walk upright on two legs.

Interestingly, our ancestors were walking on two legs far earlier than anybody had suspected. But even more interesting to ask: why did our ancestors start walking upright on two legs instead of four? Some have hypothesised that it was akin to a sophisticated chimpanzee posture used to get the best fruits high on tiny trees. The ability to walk much farther distances, to stand upright and thus absorb less heat from the Sun (at midday, near or at the Equator, only the top of the head would be exposed to direct sunlight), and to hunt and gather across much larger and potentially more lucrative territories are all things that proponents of this theory argue helped us survive. When you're standing on two legs, you can have a better view of the savannah, whether you're looking for prey or potential threats, and you can put your arms and hands to use doing other things, like tossing pebbles or making other gestures. The more tasks that could be completed with these hands, the more effective they were, which drove evolution to concentrate nerves and complex muscles in the palms and fingers.

Regardless of the specific chain of events that led to the first upright humans, we know that standing upright was crucial to their survival and success as a species. Being able to use both hands simultaneously was a tremendous benefit for the hominins, and it's possible this was a key factor in their decision to abandon four-legged locomotion. On these journeys over the arid plains, we were allowed to bring along some baggage. Yet the biggest shift didn't even happen yet. It didn't take long for these early hominids to develop from *Australopithecus* into the genus *Homo*, the lineage that eventually led to modern man, and at that point, they were already using their hands to make stone tools.

For early humans to progress, the manufacture of tools was a crucial step. This skill came from having a large brain, but it's likely that producing tools also encouraged its development. Making tools is likely hardwired into the human brain, much like the emergence of language was. The symmetrical almond-shaped or teardrop blade, honed from flint or a similar stone, was the most notable *Homo erectus* tool. Intriguingly, after our ancestors perfected the teardrop handaxe, it stayed essentially unchanged in form for close to a mind-boggling 1.8 million years!

Our current three-age archaeological system, where our physical tools evolved dramatically, was between the Stone Age, the Bronze Age, and the Iron Age. In 1836, when Danish archaeologist Christian Jürgensen Thomsen was faced with exhibiting an unrecorded collection of notably old tools and weapons, he came up with the notion of linking human history to the material from which tools were produced. To demonstrate what, in his view, was a natural progression in technology, Thomsen grouped various materials into three distinct groups: stone, bronze, and iron. A formalisation of this concept is the identification of the Stone Age (consisting of the Palaeolithic, Mesolithic and Neolithic Age), Bronze Age, and Iron Age.

Before the Stone Age, humans used simple tools, such as sticks and grass, to fish for termites in trees and holes. About 2.6 million years ago, the Old Stone Age (the Palaeolithic) began with the invention of basic stone tools, such as hammerstones, stone cores, and sharp rock flakes. The bow and arrow was the Upper Paleolithic period's second

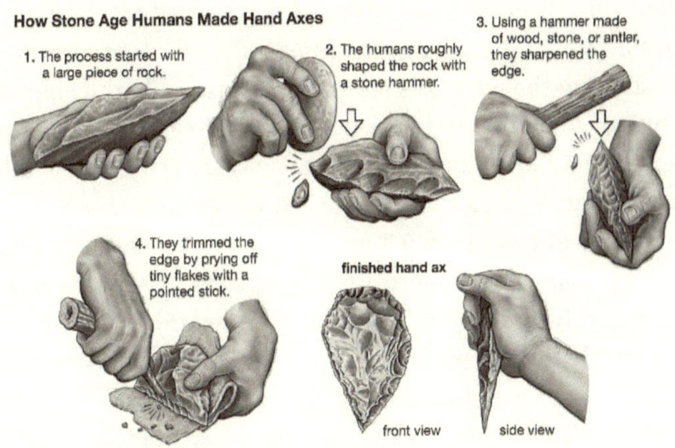

6) Paleolithic hand axes were shaped like teardrops and had two sharpened edges that met in a point. In one approach, the edges were roughed up by chipping away larger flakes with a hammer and then sharpened by chipping away smaller flakes. A sharp stick was then used to peel off small stone flakes.

great mechanical achievement, and it proved even more useful than the spear-thrower in putting more distance between the hunter and the prey. It was truly a crucial piece of equipment for Stone Age hunters. Around 20,000 years ago, in the Middle Stone Age (the Mesolithic), the pace of innovations accelerated, and humans began experimenting with various raw materials, such as bones, ivory and antlers. The latter era—starting around 10,000 years ago—is known as the New Stone Age (the Neolithic) since it was during this time that a new technique of stoneworking became widespread, leading to the general use of ground and generally polished rock tools, most notably axes.

It was quickly discovered that the contemporary style of swinging an axe with shoulder and weight behind the blade to deliver long and strong strikes proved terrible when using these sharpened instruments in a forest, either destroying the edge or shattering the blade. Short, chipping strokes should be used for proper handling, with the body action limited to the elbows and wrists. Strong Neolithic woodcutters were able to cut down an oak tree with a diameter of more than 0.3 metres in half an hour and a pine with a diameter of more than 0.6 metres in less than 20 minutes. In under four hours, three Neolithic woodcutters were able to clear an area of silver birch woodland covering 0.05 hectares (which is the equivalent of half an NBA basketball field). More than a hundred trees could be cut using just the original (ancient) sharpening of a single axhead. It was determined that the Neolithic people, armed with ground flint axes, had little trouble creating sizable clearings in the forest for agricultural reasons.

About 3,300 BCE, the protracted Stone Age came to an end with the discovery of soft, malleable, and pliable materials that did not respond to hammer blows by flaking or fracturing but instead altered their forms without breaking. Gold and silver, two of the earliest metals to garner interest among humans, were too soft to be used as tools. Natural copper and meteoric iron were the first metals of worth for toolmaking. However, they were difficult to obtain but had the ability to serve numerous functions in addition to the traditional ones.

The Bronze Age was the third stage of a major development of our tools and included axes, swords, knives, spearheads, helmets, and countless other valuable and dangerous items. Neither the Americas nor much of the Pacific Islands nor Australia are included in the three-age system since there was no Bronze Age until European explorers delivered Iron Age technology to the locals. It was true that the New World existed at the time of Columbus's first voyage and that the Stone Age was still very present in certain isolated parts of Australia and South America. Despite these caveats, the Stone-Bronze-Iron cycle is nevertheless an important notion in the development of early tools. After the Bronze Age, we moved on to the Iron Age, which started around 1,000 BCE and laid the foundation for our current tools.

However, coming back to the advantages of standing upright on two legs, there were also significant drawbacks. For millions of years, our primate ancestors' skeletons evolved to walk on four legs and had tiny skulls. It was difficult to adjust to an upright position, especially since the skeleton had to support an extra-large cranium. Therefore, humans paid with back pain and sore necks for their great vision and creative hands.

Women paid considerably extra since an upright posture demanded narrower hips, which constrained the delivery canal (and this was just as newborns' heads developed to become significantly larger).[2] Death during childbirth became a major concern for human females. Thus, women who gave birth early had a greater chance of surviving since the infant's brain and skull were still relatively tiny and flexible. As a result of evolutionary pressures, natural selection favoured earlier birth. Indeed, we can still see this adaptation of many thousands of generations ago: humans are born prematurely in comparison to other animals, with many of their key systems still immature. While a horse may gallop soon after birth and seems fully functioning, human newborns

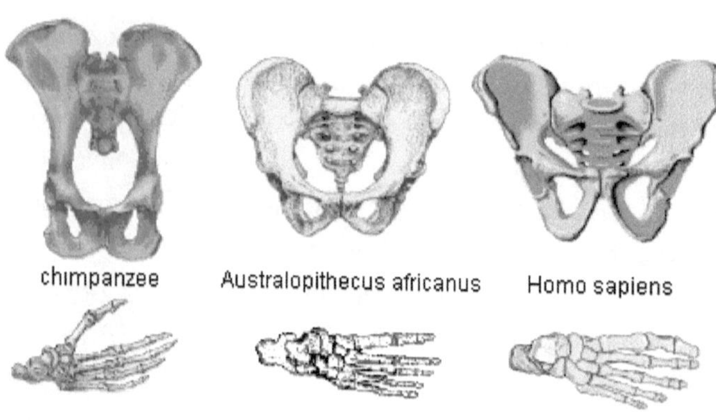

7) Comparison of pelvis and foot bones of chimpanzee, *Australopithecus africanus* and *Homo sapiens*. The relatively narrow birth canal of *Homo sapiens* is in part due to the necessity for effective balance in bipedal movement.

are immobile for many years, completely reliant on their elders for nourishment, education, and safety.[0]

For a better understanding, let's have a brief peek at a baby chimpanzee. It will have had a somewhat shorter duration in the womb than a human baby, about 230 to 240 days. Although the newborn chimpanzee is somewhat helpless shortly after birth, within 24 hours, its instincts enable it to crawl quite efficiently, and after just two days, it will climb onto its mother to suckle and be strong enough to cope without any assistance. It would take a human infant, which has just a little longer gestation period, maybe nine months or so to attain an equivalent degree of strength and control. In reality, it's quite unlikely that a newborn infant would survive for more than 48 hours if it were abandoned. A human infant's eyesight is quite weak, and they experience a hazy environment without much clarity. Also, babies have utterly underdeveloped motor skills: they can just about grab an item, they can rummage about for a nipple against their mother's bosom, and, as every new parent swiftly realises, they are fairly proficient at wailing (but that is about the end of their evident ability). Almost all of the cognitive, muscular, and visual growth that a human infant will experience in its first year has already occurred in a chimpanzee infant after several days. If humans were to enter the world at the same stage of development as a chimp, pregnancy would take as long as 18 months. Compared with our lowly counterparts at birth, you could arguably say that we are almost embryonic!

Since the majority of a human's growth happens after birth, it stands to reason that a young child will rely heavily on his or her parents for quite some time. The time and energy it takes for parents to protect and raise their offspring is significant. Humans must have a compelling evolutionary rationale for being so dependent and defenceless while they are young. What possible benefits could justify putting your trust in other people for so long? Look no farther than the larger, better human brain.

This truth has significantly influenced humankind's extraordinary social capacities and its uncommon social difficulties. Raising children need ongoing assistance from family members, friends, and neighbours. Because it takes a whole tribe to raise a human, evolution favoured those who could create strong social bonds. Furthermore, because humans are born immature, they may be educated and socialised significantly more effectively than any other species. The majority of animals emerge from the womb like porcelain (any attempts to remould them will just scrape or damage them). On the other hand, human infants emerge from the womb like molten glass, capable of being spun, stretched, and sculpted with a remarkable degree of freedom.[0]

However, we believe that a large brain, the ability to use tools, high cognitive capacities, and sophisticated social systems are tremendous benefits. Yet, despite having all of these different qualities and advances, humans remained fragile and feeble creatures for more than two million years. Thus, despite their sharp stone tools and great minds, humans a million years ago lived in continuous terror of "superior" predators, seldom hunted large game, and survived mostly by picking up insects, collecting plants, and eating the carrion left behind by mighty carnivores.

As a result, our physical tools were critical to our survival. Cracking open bones to access the marrow was a typical use of early stone tools. Many scholars believe this was our initial niche, and their explanation is simple. Assume you lived a million years ago and witnessed a bunch of mighty lions hunt down and consume a buffalo. You will probably patiently wait for them to finish. However, even after they are full and leave it's still not your turn because other dangerous carnivores, such as hyenas and jackals, would also scavenge the leftovers (and you probably wouldn't dare to interfere with them either). Only once they had gone would you and your group dare approach the carcass and delve into the delicious leftovers.

This is crucial to comprehending our history and psyche. Humans were securely in the centre of the food chain until relatively recently. Humans have hunted lesser species and gathered what they could from the start of time, while being hunted by larger predators. Only around 400,000 years ago did numerous human species begin to hunt huge animals on a regular basis, and humans only ascended to the top of the food chain in the last 100,000 years (with the advent of *Homo sapiens*). Natural selection favoured clever *Homo sapiens* because they would hunt better, survive for longer, reproduce more successfully and are more likely to ensure the survival of their offspring.

However, that stunning climb from the middle to the top of the food chain had far-reaching consequences. Do you recall the coevolutionary process from the previous chapters? Other top-of-the-pyramid creatures, such as lions and crocodiles, gradually evolved into that position over many millions of years, allowing the ecosystem to mature, balance, and adapt, preventing lions and crocodiles from causing too much havoc. As lions got more lethal, evolutionary pressures forced gazelles and antelopes to become faster. On the other hand, humans ascended to the top of the food chain so abruptly that the ecosystem was not given enough time to respond.

Furthermore, humans, according to Harari, failed to adapt themselves. We have lately been one of the savannah's outcasts and underdogs for many millions of years, and we are full of insecurities, anxieties and fear about our position and status, making us extraordinarily vicious and dangerous. It's in our DNA, and many historical disasters, from horrific fatal wars to ecological disasters, have arisen from this "sudden" leap.[0]

Archaic life in the savannah was short and precarious. The food supply and other resources, like clothing and shelter, were unreliable and of varying quality, and there were several life-threatening environmental hazards.

Humans' strength resided in their intellect as feeble, furless bipeds. The thoughts and feelings that benefited them best were implanted into their minds, and they still influence many aspects of human behaviour today. Emotions take precedence over logic. Those who survived in an uncertain world always had their emotional radar, sometimes known as intuition or instinct. And, whether faced with wild predators or looming natural disasters, Stone Age humans learned to rely on their emotional radar above everything else.

That reliance on instinct saved human lives by allowing humans with keen instincts to reproduce. So, just like any other animal, emotions are the primary filter for all information acquired by humans. And as we shall see in the next chapter, our modern challenges may differ from those of archaic Stone Age hunter-gatherers, but our hardwiring does not.

5. HUMAN INSTINCT

A few million years ago, our hominid ancestors climbed down from the trees in the thinning for us to try their luck on the savannah. As the Ice Age advanced, they were forced to adapt to less-than-ideal conditions, such as a lack of water and food and poor protection from predators. Over the course of 200,000 years, the ape-men would face out against creatures who were quicker, stronger, tougher, more venomous, and fundamentally more adapted to the violence, mayhem, and weather of savannah life.

In the beginning, as *Australopithecus*, our brains were about the size of a chimpanzee's, but they grew by a factor of three over the course of three million years while we lived on the savannah. Having a brain and the mind it houses seems to have been our secret weapon and the answer to our survival dilemma. The structure of their minds started to get more and more intricate, and a more complex mind developed in tandem with the remarkable increase in the total number of brain cells (to the present level of nearly 100 billion nerve cells). Along with our phenomenal advances in intelligence, emotion, and reasoning, humans also continued to develop a wide variety of instincts.[0]

Tools were something we learnt to make and utilise. In the process of exploring its potential, we uncovered fire and its many applications. Our curiosity for the world at large led us to go off on an adventure. It was only when we started having conversations with one another that communal living reached its full potential in terms of complexity and achievement. Hungry bands of nomads may band together, share what little they have, and figure out where to get the most delicious and nutritious food and water. Cohesive and emotional ties of collaboration and kinship made it feasible for humans to form larger communities. The growing complexity and the division of labour made it possible for humanity to establish ourselves, create complex societies, and give birth to a wide range of artistic and intellectual pursuits.

Our Stone Age Mind

Yet, we are not biologically suited to a life of anonymity in big cities, constant stress, junk food addiction, or broken social bonds. Whoever thought up nuclear weapons clearly didn't consider how quickly we can band together to defeat our enemies or how readily we can turn violence on them. Separating families was an unfortunate byproduct of the quest for prosperity, and it's not always possible to satisfy all of our deepest emotional wants and desires. As a tightly knit and mutually reliant community, we were accustomed to the rumours and backstabbing that always ensued.

Therefore, there is a conflict between our Stone Age instincts and the pressures of modern life. As a species, we are forced to carry the mentality of the savannah and the pre-hominid ancestors' hereditary features through life.

These instincts represent the inborn aspects of human behaviour. Professor Robert Winston, the author of *Human Instinct*,[0] states that human nature, including our capacity for action, desire, reason, and behaviour, is largely determined by our genetic make-up, while our uniquely human instincts were refined in the harsh environment of the savannah. As we observed about DNA, beneath this expanse of code lays the formula for the evolution of the human body. There are slight differences between people, yet our molecular code is quite similar (that is why it makes sense to call it "the human genome"). One person in a thousand will have a slightly different letter pattern. The physiology of each person is unique, as are their hormone levels, their propensity to get cancer, and the colour of their eyes.

Fundamental facts underlying gene-centred evolution underpin these material facts. We have learned a lot about natural selection over the past 50 years or more, and the answers it provides are rather convincing because its logic yields extremely powerful explanations. The concept centres on the concept of the "selfish" gene.[0] As we've seen in the first part of this book, this "selfish" gene has a huge impact on the past and future of human psychology.

Human behaviour, like a speck of dust, is influenced by a variety of factors. A variety of cognitive, biological, and cultural influences exert pressure and pull on us. Some of them may oppose one another, while some may pull in the same direction. It is entirely possible that two instinctual tendencies may act at odds with each other. We're being pulled in one direction while being pushed in another, and it's our job to figure out why. According to the principles of chaos theory, predicting the path of the dust particle in advance is next to impossible.

The complexity and adaptability of the human intellect are greater. Although our genes have a significant impact on who we are, they do not enslave us. It's quite hard to separate the adaptations from everything else. We need to separate the savannah's signals from the noise of civilisation. Some aspects of human behaviour appear to be universal, but this does not prove that they are genetically determined. Despite the fact that, as pointed out by American philosopher, writer, and cognitive scientist Daniel Dennett, "all societies who employ spears throw them pointy-end-first, this does not entail that there is a pointy-end-first gene present in all humans."

When infants first open their eyes, they experience a whirlwind of activity in the brain as they begin to experience the presence of the world beyond their bodies. An infant is born with a set of survival mechanisms that are activated one by one as they develop.

A newborn can show us how human instinct (a genetic programme which will have remained largely unchanged for tens if not hundreds of thousands of years) drives us to explore the world, how we try to comprehend and make sense of the objects and people within it, how we form relationships with our family members and, most importantly, how we ensure our own survival. It's crucial to remember that the purpose of our genes is the same for a newborn baby as it is for an adult hunter: survival. As we've seen, despite their enormous brain cases at birth in comparison to the rest of their bodies, newborn infants are incredibly immature and helpless. A newborn's strong instincts kick in and push it to convince its parents to take care of its basic needs, like providing food and shelter and keeping a watch out for predators. From the very first minute of a baby's life, instinct can be heard ringing through the hospital halls (the baby's scream is a key signal, a type of SOS, warning the mother to take care of the infant). For this reason, a newborn's cries serve as a form of early self-defence. Despite its annoyance to some, crying is a baby's go-to response to every distressing sensation, be it cold, discomfort, pain, or hunger.

There are several ways in which human infants interact with their parents and the world around them besides crying and sniffing things. Babies begin mimicking adults' facial expressions very immediately after birth, will begin to grin when they see an adult smiling, and mimic their caregivers' facial expressions. So if an adult frowns, the child may start frowning too.

Interestingly, studies also demonstrate the inverse to be correct. Babies, especially very young ones, can often decipher our feelings from the emotions on our faces nearly as well as we can decipher theirs. If mom is showing signs of fear, whether it be from an unwanted visitor, a deadly predator, or a precarious cliffside, the baby will take note. They are able to pick up on warning signs in an adult's voice inflexion as well. In a cruel yet interesting experiment, women were asked to smile at their newborns while speaking to them in a terrified manner. There is no doubt that the newborns were able to distinguish these competing signals, since the tiny mites became quite anxious and confused as a result.

In this way, human children have developed a second almost universal inclination that serves them well in the outside world. While games like hide-and-seek and tag may seem pointless, studies have demonstrated that play serves a crucial developmental and social function in children. It appears that one of evolution's solutions to the fact that our larger brains need time and experiences to develop in those first few years of life is to shape our emotions and understanding of life through play, so that we are as well prepared as possible for survival in the complex adult world through our experiences as children.

The Man Ghost

As a result, it is clear that the environment in which we are raised is crucial for the formation of human instincts and that the brain cannot grow without the appropriate inputs. Susan Greenfield, the author of *The Human Brain*, recounts the story of a little child who lost sight in one eye at age six. Ophthalmologists looked for physical reasons for his blindness but found none. Then it was recalled that, as a newborn, he had spent two weeks with an eye patch on to treat a minor illness. Therefore, the eye was useless since the brain circuits that should have processed the incoming signals from this eye had not matured correctly and were likely co-opted for another purpose.

Children who have been raised by wolves and other wild creatures in the wild, known as "feral children," provide a mysterious window into human nature. Although there are only a small number of known examples of feral children, and none of them have been the subject of detailed objective investigation, these children nonetheless provide insight into the ways in which social and cultural influences impact human development.

In 1920, the Reverend J. Singh, the founder of a rural orphanage in India, was told of a "Manush-Bagha", or "man-ghost", in the jungle some miles from his village. This ghostly creature was supposed to have a human body but a hideous ghost's head. The reverend's interest was sparked, so he set out on an expedition into the jungle, accompanied by several armed bodyguards. After hiking for a while, they came upon a massive white ant-mound as tall as a two-story building. It had seven big holes around it that all led to the same entrance in the middle of the mound. The Reverend Singh and his companions staked out the ant mound and waited for the man-ghost to arrive. An adult wolf's head poked out of one of the openings as night fell, and more wolves and two cubs soon joined them. As the cubs fled, the man ghost appeared, creeping on all fours after them. The body was that of a human child, truly a kind of caveman, the head a big matted ball staring out of which, Singh could see, were piercing eyes. The man-ghost took off running towards the bush, still on all fours.

Singh decided, however, to return with a larger force to destroy the ant hill and drive the occupants out using smoke bombs. An adult female wolf escaped the mound as workers began excavating it and was shot by one of Singh's men. They went deeper into the anthill and found two cubs and two "ghosts" huddled together in a corner, where they fought tooth and nail with their captors before being bundled up in blankets and taken away.

The ghosts seemed to be two females, one of whom was a newborn, with their features nearly entirely obscured by a mass of tangled, matted hair. They eventually settled down enough for Singh to feed them raw milk and water. He shorn the girls' massive hair and gave them the names Kamala and Amala, and estimated that Kamala, the older sister, was eight years old, and that Amala, the younger sister, was around 18 months old.

8) Photo of Kamala eating. In his dairy, Signh wrote "They could eat and sit on the ground squatting down... but could not stand up at all... Kamala and Amala could not walk like humans. They went on all fours... [They] used to sleep like pigs or dog pups, overlapping one another. They never slept after midnight and used to love to prowl at night fearlessly..."

We would assume the two girls were not human based on their behaviour. All over their bodies were the wounds, cuts, and boils that were signs of their survival in the wild. Because their joints were stiffened, they had to crawl about on all fours (they surely couldn't stand). It seemed like they were nocturnal (sleepy throughout the day and active at night). Singh reports that their vision appeared to have become adapted to the night, and they had the ability to see in the dark with ease, although on this point, he stretches our credulity. They, too, enjoyed the flavour of raw flesh. Kamala had a keen sense of smell and was once discovered eating the guts of a dead chicken that had been discarded outside the orphanage's walls. Instead of using their hands, they wolfed down their food by lowering their lips to the tableware. They weren't making any moves toward communication, and it would be quite some time before Singh heard them utter a single significant sound.

They slept like cats, with their legs intertwined, and urinated and defecated wherever they happened to be. Anytime, wherever, they would relieve themselves. They gradually became more trusting and started having fun, but there was still no sign of language or even basic gesture comprehension.

Both girls fell gravely ill with dysentery within a year of each other: a roundworm infestation was discovered in them. Eventually, they were so frail that they would only get up to take a sip of water or take a pill if someone handed it to them. The older of the two children, Kamala, began to show indications of improvement, but Amala passed away the next day. According to Singh, once Amala passed away, Kamala stayed with the body and would not go. When Amala's eyelids didn't budge, she reached for them and tried to stimulate them. As two tears fell, according to Singh, she cried. Kamala stayed in a corner for the next six days, refusing to interact with anyone. When Kamala ate too much, she often got sick. She avoided killing animals but would bring home the carrion she discovered, occasionally scaring away vultures in the process. She would occasionally disperse the corpses throughout the orphanage and try to hide them. She also started craving sugary treats. She learned the names of several of the infants at the orphanage and picked up a few phrases, such as "yes" and the term for "clothing." She learned to recognise colours and grew used to American culture enough to use the restroom without fear of embarrassment from the Singh family. There was no more development for her. After living with the Singhs and their foundlings for nine years, Kamala became unwell again (tuberculosis) and eventually passed away.

When it came to communication, the wolf-girls of Midnapore could never express themselves beyond communicating the most basic needs. Their story exemplifies the significance of a person's upbringing, family, and community. Face recognition, language learning, and the maturation of our emotions are just some of the cognitive tools we need to navigate the environment, and they won't just appear by themselves. It may be too late to "turn them on" once a certain time has passed.

The Wiring of the Brain

Culture emerged long before evolution had finished moulding us into what we are now, much like the development of a child is inextricably linked to the process of evolution. Professor Clifford Geertz points out that while humans have not evolved to better use aeroplanes, this is not always the case with items that were devised a few million years ago. *Homo habilis* is credited with inventing the first stone tool, a basic chopper or flake created by crushing two pebbles together. This invention may have had an impact on the development of our opposing thumb, our posture, the size of our teeth, and most crucially, our cognitive abilities.

The control centre for all our mental capacities and instincts is our brain and spinal cord. Dr Paul MacLean, a notable neuropsychologist from the United States' National Institute of Mental Health, put out the concept that the brain may be divided into three regions in the 1950s. His idea was labelled "the triune brain."

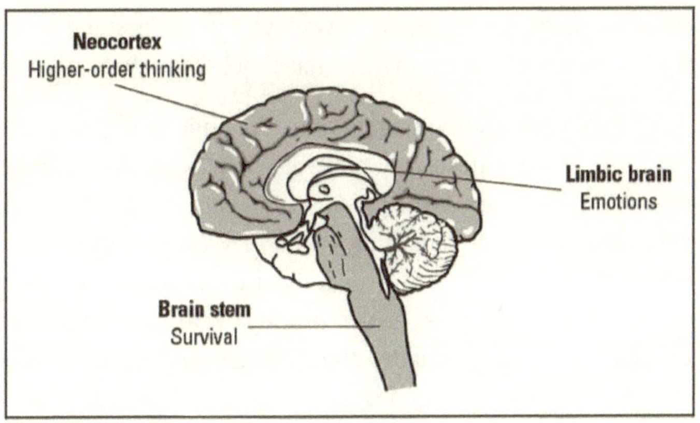

9) An illustration of MacLean's hypothesis depicting the reptilian complex (basal ganglia), the paleomammalian complex (limbic system), and the neomammalian complex (neocortex) as components gradually added to the forebrain throughout evolution.

The first, he said, was the reptilian brain, the primitive centre of the central nervous that is shared by all reptiles and is responsible for their most fundamental physiological processes, including breathing, heart rate, and digestive function. Mating, aggression, and anger are all part of this fundamental behaviour. The limbic brain, as described by MacLean, is a layer of processing power that sits atop the more primitive cortex. Your range of feelings will be severely limited without a functioning limbic system. The neocortex, the brain's third primary structure, developed as a result of the brain's continued expansion over evolutionary time. According to MacLean's theory, the neocortical brain gradually evolved to give reasoning and eventually led to human abilities like language, planning, and writing. (Although "newer" studies raise some valid criticisms of Dr MacLean's rigorous subdivision of the brain, his overall approach is reflective of our knowledge of the evolution of the mammalian brain. Of course, nowadays, our understanding of the neurology of human instinct has reached a new level).

A very well-known human instinct that you have probably experienced yourself is fear. Our hardwiring regarding fear generates an immediate and powerful emotional response, ensuring that every *Homo sapiens* pay attention and respond to a perceived catastrophe. In other words, fear pushes us to act quickly. Fear paralyses our capacity to think rationally, detect issues, and make careful judgments, because thinking takes time, and even if it was only a marginal bit for a hunted Sapiens, it just simply wasn't enough. The only viable option to survive is to either fight or flee instantly! Unfortunately, many of today's modern threats cannot be fought since no easily confrontable attacker exists (perhaps only the tax letter). And you can't run away from them either since many of them are diffuse rather than localised (you may run, but you can't hide). And although burying your head in the sand may work for ostriches, it exposes a major portion of the human body.

According to Andrew Tarantola of Gizmodo, a 2012 study conducted by academics at the University of Toronto in Canada concluded that this anxiety is not a full-fledged panic reaction. Instead, it's more like a lingering, impending fear that keeps us alert, which is precisely what our ancestors needed. This anxiety is your body's way of keeping you alert in case you need to "flee or fight" away from danger, which is arguably very ineffective against the famed feared tax letter!

Yet, even though it may sound silly, our fear of the dark is an evolutionary feature that *Homo sapiens* developed in order to avoid real-life predators lurking in the dark. Scientists think that this intrinsic dread derives from a time in human history when we were not nearly the top predators that we are now. *Homo sapiens* only became super predators with the invention of technology, which was, relatively speaking, not long ago.

Prior to the invention of modern technology, our ancestors were always on the lookout for predators trying to snack down on human flesh. To make matters worse, the majority of these predators hunted at night, when Sapiens are most vulnerable to ambush due to our weak eyesight. This means that it was extremely important for our ancestors to be safe in the middle of the night. If they didn't, they'd die. This nightly terror became instinctual over time, and we still experience it now as a type of moderate anxiety.

Even though nowadays, *Homo sapiens* don't need this fear nowadays, it's still there and perplexing. Our fear of the dark slowly morphed as early human civilisations evolved into the metropolitan area societies we have today. Only nowadays is it strange because most of us don't need to fear the dark, especially with lamps, and smartphone devices, which makes, for better or worse, darkness a choice rather than an inevitability. These traits are typically passed down through generations of distant relatives to the point that they become embedded in our psyches. Given how long humans have been around, it was only lately that this fear became nearly irrelevant for those of us who live in metropolitan areas.

Despite the fact that humanity is at its peak in terms of technology, urbanisation and lighting, you may still be triggered by this hardwiring. Envision yourself making the trip home late on a cold, damp, and foggy winter night. This day has worn you out, so you can't wait to come home, lock the door, and relax after a long day of work. As you saunter along in a state of mental neutrality, you become aware of the measured but accelerating footsteps behind you. From behind you, you can see a strange man approaching in the dim light of the roadway. He's walking a bit faster than you are, and his eyes never leave yours. No one else can save you, and the stranger can be seen on the street. The distance to the house expands dramatically. You go from feeling safe to being petrified in the blink of an eye. You feel your heart racing, your mouth getting extremely dry, and an overwhelming need to bolt out the front door and into the safety of the neighbourhood.

There's a simple explanation for your extreme apprehension. The chaos within your body is out of control. There are biological alarms and sirens sounding, and your brain and autonomic nervous system (the automatic controller of the gut, heart, arteries, and lungs) have gone into overdrive in response to the possible mugger's threat, causing a massive surge of adrenaline. The body's response to fear and aggression is lightning-fast because of the exact but complicated combinations of hormones that are released. It's like a hub of a network of nerve cells in the brain, a "fast reaction unit" ready to spring into action at the first sign of trouble. Before you have time to process what's going on, the amygdala (a little almond-shaped nucleus in the brain) has activated the fear reaction, and a biochemical cascade within you has begun, preparing your body for this potential threat. Upon detection of danger, the amygdala triggers the mobilisation of the body's physical defences in preparation for either a fight or a flight.

So, what triggers this emotional and bodily response? We weren't trained as kids to start breathing faster in dangerous situations, and we certainly can't make our hearts beat that quickly or make our bodies produce adrenaline against our will. What we're feeling is our own personal connection to our most ancient human ancestors, a reaction that hundreds of thousands of years ago almost certainly meant the difference between life and death but now, in most cases, simply serves to remind us of the remarkable fact that while we live in a very advanced modern world, we all do so with Stone Age-wired brains and bodies.

Our early human ancestors faced a hostile and perilous world. When they finally made it out of the woods and into the grassy plains of the east African savannah, they were outnumbered hundreds to one by ferocious and hungry predators. They weren't as physically powerful as giant apes or other huge land animals like lions. Neither were they as swift or nimble as, say, an antelope or a gazelle. They weren't built for flight or aquatic survival (therefore, they couldn't do either). They had no ability to see in the dark, no acute hearing to hear prey rustling in the grass hundreds of metres away, and a very primitive sense of smell. Ape-man babies were completely dependent on their caregivers and unable to care for themselves, which took attention away from more pressing survival concerns. While wandering the wide African plains in search of food, shelter, and mates, these unprotected and defenceless prototype humans had to endure the scorching heat. They might die of malnutrition or be ambushed by a lurking predator if they stayed there, or they could put themselves through the ordeal of the unknown and risk running into a terrifying beast if they wandered (and even though you weren't there at the time, you can definitely picture and assume that these prehistoric predators were terrifying!).

While the most probable threat you may experience today is a suspicious person in a local street (or perhaps the tax letter), our ancient ancestors had to face the reality of encounters with violent sabre-toothed cats and other predators (it is highly likely that predators such as these would have killed and eaten early humans on a regular basis). Therefore, in order to increase their chances of survival, all animals require a system that keeps them constantly vigilant, fearful, and able to either fight or run in reaction to potential dangers. From an evolutionary standpoint, a fearless animal has a far lower chance of making it through the harsh environment and passing on its genes. More than eight billion *Homo sapiens* currently occupy the Earth; our species has become the most successful in the history of all life on our planet. The physiological and psychological adaptations to the danger that our ancestors developed and refined must have been crucial to their survival, for we still have them within us today.

However, the amygdala frequently makes errors because our brains are hardwired to feel before they think and to feel at incredibly rapid speeds. It's possible the slithering thing you saw on the trail isn't a snake at all. And the suspicious stranger who seems to be coming toward you in the dim street may not be a threat at all. Then, maybe we don't need to be on such high alert, right? However, don't be too hard on the amygdala when it inevitably makes an error once in a while. For one very excellent reason, emotion prepares the way for conscious thought: if it weren't for that, we would all have been bitten by snakes a long time ago!

Fear served a purpose for our ancestors (and continues to serve a purpose for us now, as we remain relatively anxious creatures): it triggered their defence mechanisms, prepared their limbs for action, and spurred them to remarkable feats of survival. In extreme situations, adrenaline may provide the body with superhuman power (adrenaline does not make you stronger, but it can boost your ability to exert yourself). The rapid, automatic response that sets off the strong biochemical cascade is a classic illustration of human instinct.

It's almost as if the instinctive human aversion to snakes is latent in our brains, waiting to be activated when the time is right or when other people "teach" us what to do. The fear we feel is produced by a complex combination of hard-wired instinct and experience, which can turn on particular fear circuits in the brain.

Such "memories" might have been etched into our genes thanks to evolution's glacial speed. Let's think about the environment that early humans like *Australopithecus* and *Homo erectus* encountered every day. After the trees withered and we dispersed across the savannah, we remained here for millions of years in a climate that was likely constant. Consider the vast numbers of generations that have called it home, grown accustomed to it, and shared that familiarity with others that came after them. Introspecting oneself in the modern day may reveal a latent "remember" of the savannah based on one's DNA.

Intriguingly, an eight-year-psychological old's make-up is likely to be closer to our intrinsic, poorly socialised mind, just as their choices are closer to those of our ancestors from millions of years ago. Humans, due to the socialising influence of contemporary life and our life experiences, tend to gravitate toward areas they are familiar with or like spending time in as they age.

Because larger brains and greater intellect have contributed so greatly to human survival over evolutionary time, they have become indicators of social status. They are the neurological equivalent of the peacock's ailment, and in the contemporary world, a large brain is something the other sex may find quite appealing. This is important because being sexually attractive to the other sex increases the likelihood that a human will procreate. With our intelligence, we can make others laugh, flirt with possible partners, and even compose beautiful poetry. However, the survival of the species depends on sex considerably more than our verbal prowess alone. As we will shortly see in the next bit, we shall take a look at how our instinct for sex goes much deeper.

Intertwined Instincts

Even when our behaviour is not blatantly sexual, *Homo sapiens* are fascinated with things like money, job, attractiveness, friendships, and competitiveness. Whether or not we are aware of it, all of these facets of human existence are intertwined with sex-related impulses.

You represent one end of a very long line of sexual success stories, which is one of the major and simplest explanations for why humans have the impulse to have sex at all. Humans are hardwired for mating and procreation. Reproduction is the ultimate goal of our genetic makeup. Even while genes don't care about anything (because they aren't sentient), it's sometimes helpful to conceive about them as if they do.[1]

If your parents hadn't bonded at the precise moment, down to the minute, you wouldn't be here today. You owe your very existence to that moment. You wouldn't be here today either if the connection that developed between their parents hadn't taken place at the precise moment it did. You wouldn't be here if their parents' parents didn't do the exact same thing, and if their parents' parents' parents didn't do the same thing before them, and so on and so on.

If you trace your family tree back about eight generations, to around the same time when Charles Darwin was born, you will discover that your own existence is dependent on the successful pairings of more than 250 unique individuals. If you trace your family tree back even further, to the time of Leonardo da Vinci (1452-1519), you will find that you are descended from at least 16,384 ancestors who traded their DNA in an earnest effort to create you. This is because you are the product of a serious endeavour to make you.

Twenty generations ago, the total number of individuals who are having children on your behalf increased to 1,048,576. Your personal life depends on the committed couplings of at least 33,554,432 men and women who lived five generations before that. It is possible that more than a billion people can be connected to you through the parents and grandparents of your direct ancestors who lived 30 generations ago (this does not contain any more distant relatives such as cousins, aunts, or uncles). When you go back 64 generations, to the time of Julius Caesar, the number of individuals whose partnership was essential to your eventual existence rises to somewhere in the range of a quintillion (for reference, this a quintillion: 1,000,000,000,000,000,000).

The math must be wrong here, right? You might find it intriguing to learn that the reason is that your bloodline is not 100% pure. Even at a genetically distinct distance, you couldn't have made it this far without some incest. There is a significant potential that one of your mother's relatives had sexual relations with a distant cousin of your father several times because you have millions upon millions of ancestors. As a matter of fact, if you are dating someone who shares your race or nationality, the chances of you being related are rather strong. Most of the people you encounter in a crowded place like a bus, park, café, or any other public place are probably linked to you. Therefore, your instant response should be "Me, too!" if someone boasts that he is a member of a royal family. We have a common ancestry in the most literal and fundamental sense.

Furthermore, our similarities are remarkable. Comparing the DNA of one individual to that of another reveals an average similarity of 99.9%. For this reason, humans are categorised as a species. To quote British physicist and Nobel laureate John Sulston: "about one nucleotide base in every thousand." This remaining 0.1% is what gives each of us our unique characteristics. There has been a lot of excitement around mapping the human genome in recent years. However, "the human genome" is actually not a real thing. The human genome is entirely unique. That would mean that no one has any special qualities. Our nearly but not quite identical genomes are the product of endless recombinations within them, which contribute to the uniqueness of each of us and of our species as a whole.

For nearly all of our existence, humans shared a common ancestor with chimpanzees. Little is known about chimpanzees' ancient past, although we were probably very similar. Then, somewhere around six million years ago, a dramatic change occurred. A group of weird new creatures emerged from Africa's tropical rainforests and started roaming the wide savannah. No matter how you look at it, we humans are apes (superfamily Hominoidea). Humans evolved from an ape lineage that first diverged from gibbons (family Hylobatidae) and orangutans (genus *Pongo*), then from gorillas (genus *Gorilla*), and lastly from chimpanzees and bonobos (genus *Pan*). About six million years ago, in the late Miocene epoch, a division occurred between the human and chimpanzee-bonobo lineages. In the process of this division, "chromosome 2" was created by the merging of two other chromosomes, leaving humans with 23 pairs of chromosomes as opposed to 24 in the other apes. Many hominin species and at least two distinct genera have arisen since the hominins diverged from chimpanzees and bonobos.

However, coming back to the line of sexual success stories, the search for a sexual partner is the first crucial step in procreation. The majority of *Homo sapiens* have complex guidelines that a possible partner must pass before they are considered a good fit. It should be noted that these are idealised traits of a partner, and that ideals are naturally tempered with experience. And yet, the term "future reproductive value" is a fancy way to describe the potential number of children a woman can have in the future. It stands to reason that a youthful, healthy, and fertile woman would have a higher probability of having multiple children who would go on to reproduce themselves successfully. Undoubtedly, natural selection has favoured males that actively pursue healthy, fertile females over the course of hundreds of thousands of years.[0]

However, males cannot judge a woman's fertility based on her appearance alone and must instead hunt for further clues. While markers of overall health, like clear skin, shiny hair, and full lips, may seem like reasonable indicators of general health, they don't tell us anything about a woman's fertility. But youth is undeniably one of the most trustworthy indicators, and people of both sexes are surprisingly competent at guessing a person's age from a single glance.

Although it's true that various men and different cultures are drawn to different sorts of female "indicators", there does seem to be one common attraction: it has been found that males prefer partners with larger hip measurements than their waistlines. It's possible that a woman's waist-to-hip ratio might provide a male insight into her overall health. Adipose fat, formerly believed to be crucial for survival for both mother and child in times of scarcity, is stored on women's thighs, buttocks, and breasts during puberty due to the effects of oestrogen. Women will stop ovulating when their body fat percentage falls below a particular threshold. Therefore, since there was never any certainty of favourable "nutritious" times, it seems reasonable that over the course of millions of years, natural selection favoured males who mated with female partners with larger hip measurements (those females were less likely to starve in difficult times, which also gave the children a better chance of survival).[2]

Testosterone is an additional factor in future reproductive value. Although testosterone is commonly associated with men, females also have trace levels of the hormone circulating in their bodies. Over the course of millions of years, it's likely that the men of our hominid ancestry developed effective biological systems to fulfil these duties. Testosterone increases muscle mass and strength, both of which are useful while fending off dangerous predators. But everyday testosterone production has a major effect on many aspects of a man's behaviour, not the least of which is sex drive, in addition to its effects on muscles. There's little doubt that testosterone is to blame for men's stronger sex desires, which may have developed in tandem with their need to have as many children as possible and spread their genes.

A lot of mysteries surround our ancestors' savannah life. It's hard to say if males were essential for hunting, gathering, or raising children, but we do know that human infants are more helpless and require more attention than those of any other species. The new parents probably spent a lot of time feeding and nursing the infant (In most cases, breastfeeding was not only the greatest option but also the only one).

It's tempting to look at our modern lives, our drives, our desires, our hopes, and our troubles to find a nice, simple evolutionary explanation for them all. But we need to be careful. Clifford Geertz proposes that the study of mankind should not entail the simplification of complex explanations. Instead, he suggests that it should include exchanging straightforward pictures for more complicated ones while making an effort to keep the same level of clarity that the simpler images had. The strongest force comes from our evolutionary history. Nonetheless, cultural factors will always serve as a prism through which the hereditary component of human behaviour may be understood. However, as we shall see in the following chapter, the development of our complex brain, pre-programmed reflexes, extensive knowledge and sophisticated range of intelligence was not an easy and straightforward process. To put it mildly: our type of cognitive advancement is extraordinarily unusual.

6. THE TREE OF KNOWLEDGE

The Cognitive Revolution is defined by the advent of new ways of thinking and communication between approximately 30,000 and 70,000 years ago. Aside from the invention of cooking, another widely accepted idea contends that unintentional genetic mutations (also known as the Tree of Knowledge mutations) changed the inner wiring of our brains, enabling them to think in unprecedentedly intelligent ways. But what is so special about our abstract thinking capabilities? What is it that makes us so clever, creative and dominant? And most importantly, what is intelligence?

Cognitive Wiring

In short, intelligence is the ability to gather knowledge, be creative, learn, engage in critical thinking, or form strategies. However, consciousness is also related, which makes it very complicated to have an exact picture. Senses are the first important part of intelligence and help us navigate and react appropriately. The five human senses are sight, hearing, smell, taste, and touch.

Humans have an inquisitive nature. Regarding our evolution, it makes sense that humans are curious about the world around them. The first archaic Sapiens wandered the savannah in search of food. After hours without any success, gasping for food, suddenly something super striking is in sight: a tree with colourful hanging wild fruit next to a massive rock by a meandering river. Because of the striking colour, they slowly stroll to the tree to carefully and attentively look at what it is. After the fruit has been touched, sniffed and approved, the fruit is picked. As any predator could be nearby, the fruit is eaten quickly.

While this is obviously a very successful find, it would be extremely helpful if these archaic Sapiens could remember certain important features. If you were one of these curious archaic Sapiens, how would you recognise the same delicious fruit next time? The colour? Maybe the smell? Or perhaps the huge rock near the meandering river? However, this crucial information about where the fruit was and what it smelled and looked like has practically no meaning and value if they can't save it, which is why memory is the second essential part of intelligence.

Memory is the ability to save information and recall it to have a basic foundation of knowledge, which is very helpful for an individual not to start from scratch whenever it perceives something relevant.[3] In the case of these archaic Sapiens, all the alarm bells will ring if they encounter a similar colour, odour, or large rock near a river again. Without this memory, they will forever reencounter the fruit for "the first time."

Another creature with extraordinary memory abilities is the raccoon. In a study, raccoons were given numerous secured, transparent boxes with various locks, like latches, bolts, plugs and push bars. A lot of food (mainly leftovers from birds) was presented in these transparent boxes, which ensured that the raccoons did everything they could to get the treasure. The curious cunning racoons needed a total of 107 attempts to open the boxes, which is already fairly impressive. However, when the same raccoons were presented with the same boxes again more than a year later, they recalled how to open them and, surprisingly, did so in fewer attempts. The test of the raccoon's memory abilities demonstrates that once learned, the ability to undo simple fastenings remains virtually unchanged during periods of no practice of more than a year. In the instance of the "combination" locks, memory was insufficient after 286 days, but relearning was quick, with only 24 attempts required to obtain a facility that previously required 107.[1]

So, in addition to retaining information, it must also be organised into a series of ideas or actions that may be adjusted and modified, which is referred to as learning. Learning is a process that results in change as a result of experience, increasing the potential for increased performance and future learning.[3] Animals can learn and adapt to changing environments. If an animal's environment changes, it may no longer yield results, requiring it to adjust its behaviour.

Bumblebees are fascinating animals and have an excellent gift for advanced learning skills. Subsequently, in a study, bumblebees had to move a displaced ball to the stipulated location to obtain a reward. Bumblebees that observed a demonstration of the technique from a live example learned the task more efficiently than bumblebees observing a "ghost" demonstration (ball moved via magnet) or without demonstration. Rather than mimicking the demonstrations in which the balls were transported over vast distances, the bumblebees solved the problem more quickly by using the ball nearest to the target, even if it was a different colour from the previously observed ball.[3] Such unprecedented cognitive flexibility hints that entirely novel behaviours could emerge relatively swiftly in species whose lifestyle demands advanced learning abilities.

So with these three fundamental extraordinary cognitive abilities, creatures can memorise all kinds of associations, learn quickly, navigate and react appropriately through their senses, which enables them to act in intelligent ways. However, creatures need even more unprecedented cognitive flexibility to solve more challenging problems. One of the fanciest cognitive abilities is creativity. Being creative means developing something new from unrelated things to creating something useful. In the context of intelligence, this means making new, unusual connections by combining input with memories and skills to develop an exceptional solution to a specific problem.[3]

In the case of our archaic fruit-picking Sapiens, this would mean that perhaps the next time they encounter wild sweet fruit from the same tree, they come up with something "creative" so as not to have to climb the tree and thus obtain the fruit faster and safer. Perhaps one of these Sapiens suddenly remembers that when it mounted the same tree last time and "accidentally" shook the branches around the fruit violently, leaves fell off. The falling of the leaves, which can be seen as an unrelated thing, can eventually be turned into something useful by making an unusual connection between the vigorous shaking of the branch, the falling of the leaves and the tempting hanging fruit.

Another worthy example of developing something new from unrelated things to creating something useful is, perhaps yet again, another raccoon study. This test is based on the ancient Greek storyteller Aesop's classic tale. In the story *The Crow and the Pitcher*, a thirsty corvid goes through this exam in order to drink. Raccoons had to solve the identical problem in this study. Researchers tested raccoons' cognitive flexibility and creativity by placing them next to a water tank with a floating marshmallow in the water. The twist, however, was that the water tank was too small to get the floating reward. However, the raccoons were given different attributes to solve the problem. Eventually, raccoons learned to drop rocks into a water tube to retrieve a floating reward. When faced with functionally different choices, the raccoons will either choose the correct option at the start of the trials, exhibiting causal comprehension, or will learn to choose the correct option over time. One raccoon went to the top of the water-filled tower and rocked back and forth until the structure tipped. And the two raccoons who worked with floating objects discovered that by repeatedly pushing down on floating balls, they could create waves that splashed marshmallow bits within reach.

A notable, unique and remarkable example of intelligence is the arts of *Physarum polycephalum*, an acellular slime mould (popularly known as "the Blob"). *Physarum polycephalum* has been shown to exhibit characteristics similar to those of unicellular creatures and eusocial insects. For example, a Japanese and Hungarian researchers team has demonstrated that the Blob can solve the shortest path problem. When grown in a maze with oatmeal in two places, the Blob retreats all over the maze except for the shortest route connecting the two food sources. When given more than two food sources, the Blob appears to solve a more difficult transportation challenge. The amoeba creates efficient networks when it has more than two sources.[4]

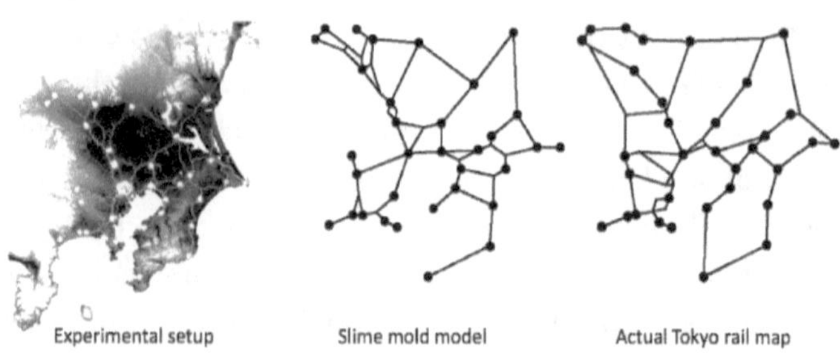

10) The *Physarum polycephalum* experiment. After around one day, the slime mould had developed a network of interconnected nutrient-transporting tubes. Its design resembled that of the rail network surrounding Tokyo.

In a 2010 study, oat flakes were distributed to represent Tokyo and 36 surrounding cities. The main oat was where Tokyo was (they inoculated it) and then let the Blob grow. First, it grew out randomly exploratory and then after about 28 hours, it reorganised itself in the most efficient way possible and reorganised the Japanese subway system more efficiently than it's designed today. Similar results have been shown based on road networks in the United Kingdom and the Iberian Peninsula (Spain and Portugal).[4]

Not only can the Blob solve these problems, but it also exhibits some form of memory. By repeatedly cold and drying the test environment of a specimen of the Blob at 60-minute intervals, Hokkaido University biophysicists found that the slime mould appears to anticipate the pattern by responding to the conditions in which they do not change the conditions for the next interval. When repeating the conditions, it would respond at 60-minute intervals and test at 30 and 90-minute intervals. These examples demonstrate cellular intelligence and are the tip of the proverbial mycelial iceberg.

However, in addition to exceptional cognitive creativity, planning is also linked to one of the fanciest parts of intelligence. Planning is the process of thinking regarding the activities required to achieve the desired goal. Planning is based on foresight, the fundamental capacity for mental time travel and involves logic and imagination to visualise the desired end result and the steps necessary to achieve that result. The evolution of forethought, or the ability to plan ahead, is supposed to have been a driving force in human evolution. A key distinction between planning and forecasting is that forecasting attempts to foretell the future, whereas planning imagines what the future may look like.

With highly instinctive behaviour and highly advanced cognitive abilities, squirrels are among the smartest and most strategically intelligent animals on this planet. Previous research has shown that squirrels react differently when caching almonds, hazelnuts, and walnuts compared to other nut species, and that squirrels can distinguish between different nut attributes such as weight and perishability. In order to include weight in the analysis, each nut was weighed and allocated a unique code. Another complicated and advanced cognitive thinking skill of squirrels is that when they feel watched, they pretend to bury nuts in order to distract their competitors from their hoarded food source. Eastern grey squirrels (*Sciurus carolinensis*), usually known as simply the grey squirrel outside of North America, hoard food in the presence of conspecifics and engage in behavioural deception by covering more empty locations where nothing has been hoarded.[5]

Squirrels have also shown they are superbly capable of remembering where they have buried nuts. In a study performed at Princeton University, grey squirrels were capable of using spatial memory to retrieve nut caches they had buried. They can remember good food sources from year to year and can also memorise the easiest and quickest route up a specific tree to get back and forth from their nests.

In his online video "Backyard Squirrel Maze," former NASA engineer Mark Rober creates intense obstacle courses built for squirrels as a deterrent to stealing food from a bird feeder. Rober is careful to note the one-way exits along the way so that they can nope out at any point. Eventually, the squirrels get through every complex obstacle after numerous trials. While the video was primarily made for entertainment purposes, it's a super fascinating and interesting example of how unbelievably advanced their cognitive abilities are.

But if these intelligent animals have it "easier" to survive when they have super sophisticated cognitive abilities, why don't all animals have such complex and advanced intelligence? The simple answer is that evolution cannot and will not want to afford these expensive, prestigious, sophisticated cognitive skills. See, some species don't need advanced cognitive skills and tricks up their sleeves if it's not necessary for their natural habitat and ruthless survival.

Squirrels are omnivores and need to defend their territories fiercely. So for these little gnawing animals, remembering where their nutritious treasure is and how to trick their rivals is very helpful for their survival. But for *Ovis aries* (sheep), this is not necessary. Sheep do not have to have advanced tricks such as squirrels. They are grazers, and the most important skill they have are social skills. Sheep are frequently thought of as unintelligent animals. However, sheep can recognise individual human and ovine faces and remember them for years. They can remember 50 other different sheep faces for over two years, as they possess similar specialised neural systems in their brains' temporal and frontal lobes to humans and have greater involvement of the right brain hemisphere. So for sheep, evolving and retaining the sophisticated cognitive abilities they will never use would just be a complete waste of energy.

However, besides planning and strategising, inventing and making practical use of physical tools is seen as the most refined complex form of cognitive intelligence.[3] As we already examined, mastering fire created the first massive difference between humans and other animals, as the strength of practically all creatures depends on their bodies (height, muscles, tooth size), whereas ours (fire) resulted in practically unlimited power when completely controlled and ignited.[0] All our physical tools from the three-age archaeological system drastically increased our net potency and strength, which grew exponentially alongside each technological development and refinement.

But in addition to *Homo sapiens*, many other social mammals have been observed engaging in tool use. A group of Indo-Pacific bottlenose dolphins (*Tursiops aduncus*) in Shark Bay uses sea sponges to protect their beaks while foraging. Sea otters (*Enhydra lutris*) will use rocks or other hard objects to dislodge food (such as abalone) and break open shellfish. Egyptian vultures (*Neophron percnopterus*) regularly carry rocks in their beaks before throwing them over ostrich eggs to break them and eat their contents. Last

11) Capuchin monkeys smash nuts and seeds between a smaller hammer stone and large anvil-like rock.

but not least, let's not forget our closest living relatives (chimpanzees, orangutans, gorillas and bonobos), which all share the ability of tool making, for they will fabricate and use long sticks to fish out termites and ants from their mounds, and even carve sharp spears to hunt for prey.

The Power of Fiction

While most species only evolved particular tools, features and intellectual sophistication, humans (particularly *Homo sapiens*) invested in a remarkably diverse range of cognitive abilities. By combining and evolving a diverse range of intellectual skills, we accidentally created another tool on top of it: communication. Of course, practically all animals interact and communicate with each other. But it's far more important to understand the consequences of our Tree of Knowledge mutations. So, what exactly is it about our language that has empowered us to conquer the entire planet? And how did our language initially arise in the first place?

Before we go deeper into the power of our language, let's explore its historical evolution. Given that language did not leave a physical trace until the invention of writing, it is logical for archaeologists to use art, decorative shells, pigments, and stone tools as proxies for the probable origin of language. So far, no art or artefacts have been discovered that are older than about 100,000 years. Michael Corballis is a notable scholar in the field of language evolution, and he has written extensively on the subject. Corballis argues that gesture communication initiated the first linguistic revolution. He thinks that early hominids communicated with one another through basic gestures, which later developed into the complicated language system we use today.

In their eloquent book *Origins of the Social Mind*,[9] evolutionary developmental psychologist Bruce J. Ellis and professor of psychology David F. Bjorklund argue that competition for food, status, power, and access to females intensifies when group size rises. Primates evolved a big neocortex in response to these stresses. But, some monkey species have developed strategies for maintaining social cohesiveness that do not need significant brain size increases. In addition, hominids were under pressure to maintain big social groupings much before brain capacity increased significantly. From 4.5 and 3.5 million years ago, East African hominids experienced an extension of their range and a proliferation of species. The subsequent epoch was characterised by intense competition and range reduction. Throughout this range expansion and contraction era, our ancestors did not improve their social structure by growing larger brains, although brain size did rise a little. Pliocene hominids probably solidified their group structure through a series of focused neurobehavioral modifications. The subordination of the vocal system to cerebral control may have been the most important of these. This assertion is supported by the fact that upright posture and complete eye contact allowed for the emergence of the earliest gestural communication amongst early hominids. As many have claimed, it is probable that early hominids relied on certain sorts of gesture communication.

Our Pliocene ancestors likely had a vocal repertoire similar to that of other primates. It was not the introduction of new sounds that contributed to social cohesiveness but rather the capacity to utilise existing sounds in new circumstances. By recalling certain calls and gestures at will, our ancestors could grab their peers' attention and negotiate the fundamentals of group interactions through chatter. The reason that the great apes did not undergo a similar evolutionary process in West Africa can be linked to the distinct group size needs in their arboreal habitat and the fact that they had not evolved bipedal stride and the resulting advances in face-to-face communication.[9]

The second linguistic revolution appears to have been mimesis (to imitate), which was extraordinarily efficient in fostering social cohesion and collaborative planning. As a result, *Homo erectus* was able to spread its territory throughout the entirety of Africa, causing the eventual extinction of all other hominid species. It also enabled the successful migration of *Homo erectus* from Africa to all of Asia, including the Middle East, China, and Indonesia. We may owe this prosperous expansion to two fundamental mechanisms.

The first was the capacity to destroy rivals, especially in Africa. To do this, *Homo erectus* must have depended on social cohesion as a basis for combat. Second, for *Homo erectus* to be able to migrate to new regions, they must have been highly adaptable. This adaptability could not have resulted from a straightforward physical change. Instead, it must have been the outcome of a general increase in cognitive ability, especially as indicated in cooperative problem-solving and adaptability.[9]

Selection for language-related proficiency is predominantly driven by mate selection, with both males and females favouring mates who can mimic, speak, and think adequately. Language may be utilised to establish the love and friendship that underpin the majority of sexual relationships, as well as to identify continuous infidelity in these relationships. Simultaneously, much of the evolutionary success of language-based love may stem from the use of language to mislead. By fabricating mental pictures of fidelity and pair bonding, males can obtain sexual favours from women even if they are not faithful. This strategy type corresponds to what we currently refer to as "a smooth talker." Much bigger benefits are achieved to those who undertake charismatic group leadership responsibilities.[9]

Still, there had yet to be another linguistic revolution to acquire contemporary linguistic capabilities. Possibilities for constructing a mental vocabulary were made possible by the advent of smooth methods for articulatory planning. By encoding words into a compact set of contrasting traits, *Homo sapiens* was able to conventionalise, learn, store, and recall a nearly infinite number of names for things. Yet, the creation of a lexicon is insufficient to generate a whole human language. Humans must also employ a method for merging words into sentences. Throughout the mimetic stage, however, *Homo erectus* was already acquiring a crude form of this fundamental combinatorial capacity to regulate plans and sequences of conventionalised acts. Consequently, the blossoming of language around 70,000 years ago resulted from a connection between the new lexical power and an earlier mimetic power. Thus the most substantial shift between 70,000 and 40,000 years ago was the emergence of the ability to manage perspective-taking in the auditory mode.[9] Corballis has proposed that the left hemisphere of the human brain has a specific language module. It is believed that this module, which is responsible for processing the sounds and syntax of language, developed significantly during this shift.

With the aid of a systematised lexicon, it would be simple for humans to identify and encode all of the significant objects, properties, and behaviours in their environment. Possessing a complete inventory of the physical world enabled the first humans to exploit various animal and plant species for more sophisticated purposes. This new lexical wealth proved especially potent when it was included into the perspective-shifting mechanism. After *Homo sapiens* were capable of creating, storing, and learning a vast vocabulary of phonologically structured forms, the subsequent steps in the evolution of language were rather straightforward.

Having discussed the origin and evolution of language, let us now explore the power of our language. The most typical response is that our language is incredibly flexible. We can string together a finite number of sounds and symbols to create an unlimited number of sentences, each with its own unique meaning. We can consume, retain, and share massive amounts of information about our surroundings as a result.[0] In the case of our fruit-picking archaic Sapiens, they can precisely describe the location with each other in superb detail and profoundly discuss whether it is safe to approach the tree.

According to a second theory, our unique language originated to convey knowledge about the world. But the most crucial message required to be conveyed concerning humanity, not gazelles or lions. According to Harari, our language emerged as a means of gossiping. According to this idea, *Homo sapiens* is fundamentally a social animal, and social cooperation is essential for our survival and reproduction. This speaking ability, which contemporary Sapiens acquired some 70,000 years ago, allowed them to chat for days on end. Having reliable information about who could be trusted enabled small bands to grow into bigger bands, allowing *Homo sapiens* to build stronger, tighter, and more complex forms of cooperation.[0]

The flexibility and chitchat theories are both valid, but what truly distinguishes human language is its capacity to send information about entities that do not exist (fiction). According to experts, only *Homo sapiens* can speak about things we have never seen, felt, or smelled. The Cognitive Revolution gave birth to legends, myths, gods, and religions. Many human species also have a contemporary vocal tract and are therefore capable of completely modern speech, such as "watch out, there's a hungry lion near the river close to the tree with colourful hanging wild fruit!" During the Cognitive Revolution, however, only *Homo sapiens* learned to say "The powerful roaring lion is the guardian god of the tree with colourful hanging wild fruit." This ability to speak about fiction is the most distinctive and unique aspect of *Homo sapiens* language, allowing us to envision things collectively rather than individually. This capacity to envision things but do so collaboratively led to greater and more flexible cooperation.[8]

Gossip, on the other hand, has its limits. Many sociological investigations and research have found that the typical maximum "natural" size of a gossip-bonded group is around 150 people. Interestingly, even now, in modern organisations, the crucial threshold is somewhere around this figure.[9] Below this level, army forces, companies, tribes, and social networks can function, relying primarily on personal information and rumours. But how did *Homo sapiens* manage to cross this vital threshold?

The secret was most likely the rise and sophistication of fiction. As we discussed, strong and complex fiction resulted not only in "imagination" but also in collective thinking and extremely flexible collaboration. Every large-scale human cooperation, whether a contemporary state, metropolis, or big ancient tribe, is based on shared stories that exist only in people's collective imagination. Innumerable separate "cultures" arose via fiction, each with its own power, intensity, reality, and laws. Culture enabled our ancestors to prosper and spread into new territories. As a consequence of the connection between culture and genes, humans have acquired various crucial adaptations, including decreased aggression, cooperative attitudes, collaborative skills, and the capacity for social learning.[7]

All other animal species exist in a dual reality due to the lack of this ability. On the one side, they are aware of physical objects outside themselves, such as forests, mountains, and waterfalls. On the other side, they are conscious of emotional feelings, such as fear, joy, and desire, that occur within them. In contrast, *Homo sapiens* exist in a reality with three dimensions. In addition to the spiritual forests, the holy waterfalls, and thoughts and feelings, the universe of *Homo sapiens* also contains deep-rooted stories about currency, gods, authorities, and commerce. As time proceeded, the dominance of gods, authorities, and commerce expanded at the cost of waterfalls, fears, and needs. The latter layer has blocked and domesticated the spiritual forests, and they have mastered how to shape our deepest fears and desires.

Consider how incredibly powerful this third layer is: Since they both believe in the same God, two Muslims, or two Christians, who have never met before, will know that the other will have somewhat the same morals and values as themselves. Therefore, without sharing any verbal contact, they know how to converse and collaborate far better than someone who is neither Muslim nor Christian. In the context of our ancient fruit-gathering Sapiens, if they see unfamiliar humans with the same culturally colourful body paint as themselves being attacked by a lion in the distance, they will undoubtedly rescue them (seeing the recognisable colourful body paint is thus a signal for the whole group that the endangered persons also "belong" to the same river guardian god: "Therefore, even without knowing the endangered persons personally, we shall help them at any cost!").

What about the Ukraine War in 2022, where hundreds of thousands of Ukrainians fight together because they all believe in the existence of the Ukrainian nation? Or

12) The Masjid al-Haram (also called the Great Mosque of Mecca, or in Arabic ٱلْمَسْجِدُ ٱلْحَرَامُ), built to enclose the Ka'bah, the holiest shrine in Islam. It can accommodate up to two million worshipers during the Hajj period (the pilgrimage to Mecca, Saudi Arabia), one of the largest annual gatherings of *Homo sapiens* in the world. Everyone in the image shares similar beliefs, morals, and values, a powerful reminder of how religion can spiritually connect *Homo sapiens*.

two people who meet on vacation on a cruise and discover that they both intensively support the same particular large sports club? Due to their shared beliefs, the soldiers who fight side by side, and the two vacationers on the cruise, will be able to communicate and collaborate more efficiently. However, none of these things actually exist apart from the stories that *Homo sapiens* have created and told one another (Nothing in this statement is intended to demean the standards and values held by people; rather, it's meant to point out, based on the findings of scientific research, that without the rise of humans, these forces, such as the nation of Ukraine, the globally famous sports club, or the sacred tree, spirits and other gods never would have been expressed, formed, or been honoured by *Homo sapiens*).

Yet, people are able to more easily estimate how another person thinks, performs, and operates with little to no effort on their part in terms of either their time or their energy. While many people think that "ancient archaic tribes" solidify their social structure by believing in spirits, holy trees, holy rocks, and the full Moon, many people fail to grasp and admit that our contemporary institutions operate on the same principles. While the capacity to build an imagined world allowed big groups of strangers to interact efficiently, it also did something else: it changed how *Homo sapiens* collaborated by altering myths and telling alternative tales.

This is very unusual because the behaviour of other social animals is mostly influenced by their genes (DNA). Significant changes in social behaviour cannot arise in the absence of genetic alterations. According to specialists, changes in social patterns, the development of new technologies, and the establishment of new habitats were caused more by genetic mutations and environmental pressure than by cultural endeavours.

This explains why humans needed hundreds of thousands of years to accomplish these enormous advances. Genetic changes culminated in the creation of a new human species known as *Homo erectus* around two million years ago. Its origin coincided with the development of a new innovative stone tool technology, which is today recognised as a distinguishing trait and attribute of this species. So, as long as *Homo erectus* did not experience any significant genetic changes, its stone tools stayed basically the same (in this case, for over two million years). In contrast, *Homo sapiens* have been able to rapidly alter their behaviour since the Cognitive Revolution, passing on new behaviours to future generations without the need for environmental or genetic modification. In other words, although the behaviour patterns of ancient humans remained the same for tens of thousands of years, Sapiens may modify their social structures, economic activities, interpersonal connections, and a range of other behaviours in a relatively short amount of time.[6]

Thus, that was the key cause for *Homo sapiens* success, and it explains why we were so inexorable over other human species. As we discussed in prior chapters, Neanderthals were physically more powerful than *Homo sapiens*. Therefore, in one-on-one combat, a Neanderthal would have likely beaten a *Homo sapiens*, comparable to two boxers battling in vastly different weight classes, with the *Homo sapiens* being a lightweight and the Neanderthal being a heavyweight. In a violent conflict between a Neanderthal and a *Homo sapiens* tribe, however, Neanderthals would likely not stand a chance against a great *Homo sapiens* cohort (Neanderthals would be relentlessly outnumbered). Neanderthals could not collaborate efficiently in large numbers because they lacked the ability to construct great fundamental fiction, nor could they change their social behaviour to quickly changing conditions (such as our cognitive revolution and strong cooperation).[6]

Thus, the mystical glue that holds together enormous numbers of individuals, families, and organisations is the fundamental difference between us and everyone else. And yet, as we'll see in the next chapter, this gooey magical force had already evolved at some subtle level in the programming of the brains of the early human species millions of years ago.

7. OUR LAST ANCESTORS

Each of the countless distinct cultures in the globe has its own stunningly original art and literature. Each individual and every country has its own history, and though we may have our differences, we all have a shared history: our human evolution. In this chapter, based upon anthropologist Stefan Milosavljevich's fascinating and interesting documentaries,[0] we'll go through the evidence for our African evolution and trace down our last common ancestor before moving on to when we left that region and stormed into the world as an invasive species.

Life and Death 3 Million Years Ago

Lucy, an *Australopithecus afarensis* and one of the most significant fossils ever discovered inhabited the African savannah around 3.5 million years ago. Lucy was born into a world that was similar to ours in many respects but profoundly different in others. This planet was untamed and completely brimming with life millions of years before our species left its mark on it. Some species would be the progenitors of present animals, whilst others would be quite different and dangerous.

The archaeological record is replete with evidence of our ancient predecessors falling prey to powerful animals when they were young *Australopithecus afarensis*. A *Paranthropus*, a strong *Australopithecus afarensis* with two huge puncture holes in the head that were most likely inflicted

13) Reconstruction of Lucy's whole body by paleoartist John Gurche. Interestingly, the excavating crew, led by American paleoanthropologist Donald Johanson and French geologist Maurice Taieb, named the skeleton "Lucy" after the Beatles song "Lucy in the Sky with Diamonds," which was played at the celebration the day she was found.

by a leopard.[1] Or the bones of (presumably) thirsty australopiths peppered with puncture marks of crocodile teeth.[2] Upon entering the water for a drink, the two individuals were undoubtedly ambushed by a crocodile lying in wait. Possibly the most miserable was the three-year-old little Taung child (an *Australopithecus africanus* toddler, dating back 2.8 million-year-old). Based on the imprints on the insides of the kid's eye sockets, it appears that the youngster was abducted from the savannah by a predatory bird, most likely an African crowned eagle, and met with a dreadful end.[3]

Lucy, however, overcame these obstacles and perils, maybe by spending a great deal of time among the trees. Although we are convinced that Lucy was bipedal due to her short and broad hip, completely extensible knee, and other physical markers, her upper body and the upper bodies of other *afarensis* hominins are more complex and appear to have retained many more ape-like characteristics. In 2012, researchers discovered the shoulder blade of Selam, a juvenile *Australopithecus afarensis*. Even though Lucy is 40% complete, her shoulder blades are unfortunately missing, depriving us of a vital piece of the anatomical picture. Selam's shoulder blades were far more similar to present juvenile gorillas than they were to modern humans, indicating that they still spent time in the trees.[6] However, it is also plausible that these were archaic characteristics that *Australopithecus afarensis* inherited from our last common ancestor with chimpanzees and that they had no practical use, evolving extremely slowly owing to a lack of significant evolutionary forces.[7]

Lucy's upbringing was likely comparable to that of current great apes: she played with other children, rolled in the grass, jumped on her siblings, observed the elders in her group, and learned the survival skills she would require. Finding food would have been the most major obstacle Lucy would have encountered on a regular basis. Analysis of the teeth of 20 *Australopithecus aferensis* specimens reveals that their diet consisted primarily of grasses that grow in arid, open savannah environments.[4] They could have consumed the grasses and seeds themselves or the far more calorically dense roots and tubers. Due to their inaccessibility, competition for these buried items is quite low, and they might have offered a nutritious meal. Our closest cousins, such as chimpanzees and gorillas, shun these foods to an almost complete degree. This shift in diet represents a major turning point in the course of our evolution, away from the forest's resources and toward the more diverse resources of open environments.[8]

Even though there was minimal competition for this food supply, their lives were not easy. The australopiths who lived one million years after Lucy exhibited seasonal dietary stress, as evidenced by their teeth.[5] Lucy certainly endured comparable struggles and spent a considerable amount of time searching for food on the savannah. Laetoli footprints, which date back to Lucy's time, have been preserved in the mud and are available for research today (a prehistoric site located in Enduleni, Tanzania). What's remarkable about these footprints is that others have recently been discovered in the same spot, indicating that hominins walked differently in the past.[16]

The extent to which australopiths were able to walk and run has become the subject of great discussion in anthropology, and these footprints will surely be analysed for indications regarding the origin of modern bipedalism. The Laetoli footprints are not just a remarkable artefact of early human existence, but also the physical footprints of our most ancient ancestors. However, more importantly, these imprints serve as a reminder that three million years ago, humankind was in its infancy and that there were likely numerous diverse groups, each with its own peculiarities, survival methods, and biological adaptations.

But as Lucy explored the savannah in an attempt to avoid starvation, she was most likely hunting for something meatier than plants. Numerous monkeys and apes utilise tools to obtain food. Consequently, it is probable that there has never been a time in the evolution of humans when we were not employing some type of rudimentary tools. However, the ability to adapt and develop tools distinguishes our lineage, hominins, from our other relatives (particularly stone tools).

Dating as far back as 2.5 million years ago (around one million years after Lucy's death) and long considered the earliest, universally acknowledged stone tool industry, the Oldowan stone tools are a significant milestone in human evolutionary history and were first made by our ancestor *Homo habilis*. However, recently discovered and excavated stone tools date to 3.3 million years ago (within Lucy's lifetime!). These have been named the Lomekwi assemblage and are the earliest stone tools ever unearthed that have been purposefully modified.[13] Animal bones, most likely those of ungulates and bovids, with cut marks dating back an astounding 3.4 million years, have also been discovered. This evidence implies that Lucy's diet might have been augmented with a significant amount of meat and that she created methods to obtain it.[9]

Nonetheless, this raises some quite intriguing questions. Initially, how did they obtain this meat? Lucy was just approximately one metre tall and a poor sprinter; therefore, it was unlikely that she was effective while hunting huge and swift creatures. Could australopithecine armies

14) Stone tool chopper belonging to the Oldowan tool industry (approximately 2.6 to 1.2 million years ago).

have intimidated faster predators off their prey? It is a distinct possibility. Other primate groups can be quite hostile, so if Lucy's life was anything like that of chimpanzees, avoiding other australopiths was likely another important obstacle. Since large predators frequently leave a great deal of flesh on the bones (not to mention the bone marrow within), Lucy might have also been scavenging abandoned animal remains for food. Perhaps australopithecines like Lucy developed stone tools to strip the bones of little slivers of flesh or to break through lengthy bones in order to reach the marrow.[15] Lucy presumably fought off other scavengers, such as vultures, rather than competing with top predators.

What was happening in Lucy's brain is the second essential question. If our interpretation of the data is accurate and hominins like Lucy were modifying stone tools, then this shows that australopiths' brains had undergone a radical change relative to those of other apes. Lucy's brain would have been comparable to that of a modern chimpanzee, but somewhat larger on average. Yet, according to research, chimpanzees are incapable or unwilling to build modified stone tools and cannot even utilise a sharp flake to open a food-containing box. Even if the difference between the two skulls is little, something profound was occurring beneath the surface. Perhaps australopiths' brains were being rewired.

The interesting Makapansgat cobble is an additional extraordinary artefact that provides a window into Lucy's mind. This pebble with natural flakes and wear patterns that make it look like a rough representation of a human face was discovered in a cave in South Africa in layers containing the bones of australopiths that date back 2.95 million years.[11] This rock is oddly located many kilometres away from its natural location, maybe as many as 30 kilometres. It is likely too large to have been carried in by a bird in its stomach, and there was no sign of floods in the cave that may have brought it there. No one can dispute that this rock awakens our "pareidolia" sense, which loves to discover faces (the tendency for perception to impose a meaningful interpretation). However, this capability is not exclusive to us. Numerous animals, such as chimpanzees, analyse faces in a manner comparable to that of humans; this ability has developed for the sake of self-preservation.[12]

Face recognition is crucial for survival since it increases the likelihood of recognising "threatening" faces, which again suggests that australopiths' brains were evolving. This is particularly intriguing because, to yet, endocasts of australopith and chimpanzee brains have revealed little structural differences, indicating that the transition must have been relatively subtle yet quite substantial.[13] Perhaps

15) Pebble from Makapansgat transported to a cave in South Africa where it was deposited roughly three million years ago.

the origins of all of our artistic expression and creativity can be traced back to Lucy's simple moments of curiosity. Lucy was not only an upright-walking ape; rather, she was taking her first hesitant steps toward becoming an actual human being.

Possibly one of the most significant moments in Lucy's life was giving birth. This was likely challenging for her. As previously mentioned, the combination of upright walking, which restricted our hips, and a gradually expanding brain began to overburden pregnant women. In order for australopith infants to pass through the birth canal, they would have had to spin slightly. As a result, giving birth was definitely more difficult than it had been previously in our development and may have required a greater degree of cooperation.[14] Lucy, unlike chimpanzees, may have received assistance from others, possibly her mother, in bringing new life into the world.

Typically, when we imagine archaic ancestors cooperating more, we envision them hunting together. However, this is simply one manner in which humans collaborate, and the growing difficulty of giving birth likely played a crucial role in the development of our complex social networks. The extent to which Lucy's sexual partner contributed to the upbringing of her children is the topic of great discussion. The problem concerns the size disparities between males and females, a phenomenon known as "sexual dimorphism." Species of primates with a high degree of sexual dimorphism tend to live in communities in which one alpha male attempts to have sex with all females and plays a minor part in childrearing.

This gap in physical traits between the sexes is surprisingly indicative of a species' sexual behaviour. Sexual dimorphism is most evident in the scientific sense in animals that adopt the harem system, in which one alpha male mates with many females. A considerable number of female mountain gorillas will congregate around the dominant males, and when there are many available females, there will be severe physical competition for mating rights. These masculine battles favoured the larger and stronger gorillas, resulting in an evolutionary pressure for males to become larger. Size counts when it comes to real physical force, but it is by no means adequate. It turns out that several of the characteristics of animals that we may have assumed evolved as a defence against predators are often employed to ward off rival males in the quest for sex. Deer antlers are an excellent instance of this.

Approximately 5% of the adult males are responsible for siring the bulk of elephant seal pups. They attempt to knock each other to the ground, similar to sumo wrestlers, in order to win a match. The victorious male "gets" the female and the ultimate prize of siring kids and transmitting on his genes, while the losers are forced to wait in the wings, offspring-less. In this case, it comes down to "survival of the fattest," don't you think?

Then what of humans, who share so many genes with other species? Among humans, men are generally bigger than women. This basic statistic shows an intriguing fact about our hominid ancestors. Our prehistoric male ancestors likely had to compete for mating rights, indicating that early human communities frequently engaged in "polygyny," in which a percentage of the males monopolised all the females. We lived in a human version of the harem system, although not to the same extent as elephant seals or gorillas, assuming all else is equal and assuming the concept is accurate. However, it is probable that males engaged in roughhousing in order to get access to women. Unavoidably, even today, men continue to compete for females.

In contrast to other primates, however, humans lack a large degree of sexual dimorphism, and males are typically actively involved in raising children. While several studies have discovered that *Australopithecus afarensis* exhibited a significant degree of dimorphism, others have claimed that the degree of dimorphism is comparable to that of *Homo sapiens* and, consequently, that australopithecines men were also active in family life.[17] In line with this theory, australopithecines appear to have evolved more slowly and had longer childhoods than our contemporary ape relatives, maybe because they had to learn how to use stone tools and live slightly more complicated lifestyles.[13]

Therefore, further funding was likely required from a third party, which may imply that it was the father. However, all of these predictions depend on reconstructions because of the lack of properly preserved male and female australopiths. Not only are the specimens we have fragmentary, but they are also separated by hundreds of thousands of years, which is plenty of time for a species to evolve. Regardless of all the twists and turns in Lucy's life, she eventually passed away. On the basis of the fractures on her bones, some have hypothesised that she perished from a fall, while others claim that these fractures may have been caused by huge animals.[15]

Her family likely lamented her passing regardless of the circumstance. Humans, like many other primate species, such as chimpanzees, grieve and mourn the dead. As evidence implies that australopiths may have developed even tighter ties, we can only imagine how painful it must have been for them to see Lucy's death. Unbeknownst to them, more than three million years later, we would be enthralled by the stories Lucy can tell not just about herself, but also about the origins of everyone alive now.

Muddle in the Middle

Typically, when we discuss our last common ancestor, we talk about the earliest hominins before australopiths. However, let's now concentrate on the evolution of *Homo sapiens*, our most recent common ancestor and our closest relatives, such as Neanderthals and Denisovans. Who were they? When and where were they live? And where have they gone?

Homo erectus was likely the first hominid to leave Africa some 1.8 million years ago, scattering archaic human groups over Africa, Asia, and Europe. They all went on to spend their lives, developing and adapting to their respective climates, frequently meeting to exchange DNA and thoughts. However, it is inherently difficult to identify our direct ancestors and untangle them from this complex web of ties. The last common ancestor existed at least 530,000 years ago, according to genetic analyses of Neanderthal remains from Belgium, France, Croatia, and Russia. However, that is only the bare minimum. It may have been as long back as 700,000–800,000 or possibly a million years ago.

Therefore, we have a span of around 300,000 years in which they may have lived without knowing who they were. This time period is also known as the "Muddle in the Middle." The name derives from the fact that we know how *Homo erectus* evolved from Neanderthals to *Homo sapiens* to Denisovans, but it is a complete muddle when attempting to determine who existed between these two ancient human groups. Researchers have a difficult time grasping the links between the fossils we've discovered thus far. Additionally, we have not discovered many fossils from this key age. Therefore, it is difficult to address these excellent questions while just working with a few specimens.

The location of La Sima de los Huesos in the Sierra de Atapuerca mountain range in north-central Spain has been a crucial mystery in this argument. This location was inhabited by a population of hominins enjoying a typical Paleolithic lifestyle around 430,000 years ago. They probably lived in

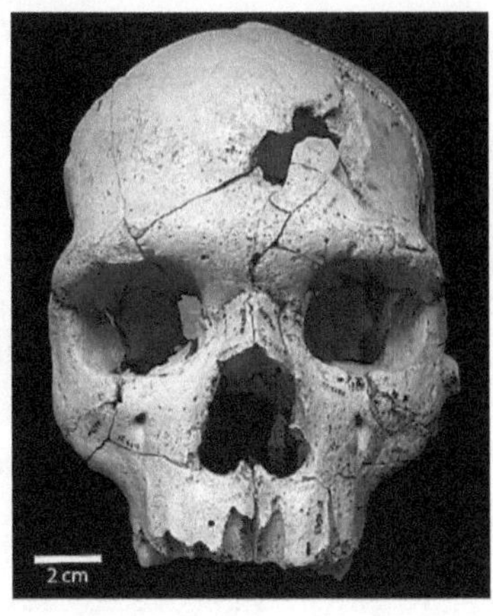

16) Cranium 17 (Cr-17), one of the skulls found at the Sima de los Huesos site, is a 430,000-year-old, nearly complete specimen made of 52 bone fragments preserving the whole facial skeleton. It belonged to a young adult and has two severe injuries above the left eye on the frontal bone.

family groupings, hunted enormous wildlife such as rhinos and horses with crude wooden spears, and may have brought their dead to this cave on purpose. There are little indications of habitation at the site, suggesting that no one lived there. We also do not observe many signs of carnivore activity, despite the fact that 28 hominins were buried there. It is probable that these ancient hominids moved their deceased to this location on purpose, posing several intriguing issues about their intelligence, culture, and way of life.[19]

These fascinating specimens have been named *Homo heidelbergensis*. Numerous specimens have been designated as *Homo heidelbergensis* due to the uncertainty surrounding the connections between the hominins. Whether they are, in fact, all the same species and should be grouped is a subject of intense discussion.

Was this group of hominins, however, our ancestor? It appears not. Genetic testing confirmed that they were early or proto-Neanderthals, living just after the break between humans and Neanderthals and so firmly on their branch of our family tree. Based on the similarities between Neanderthal and skulls from the La Sima de los Huesos site, a number of anthropologists had already anticipated this.[18] They are, therefore, not a direct ancestor of *Homo sapiens*, despite their significance. However, let's remain in the Sierra de Atapuerca and go back approximately 850,000 years.

In 1994, researchers uncovered the skeletal remains of a number of individuals who had evidently been victims of cannibalism, a practice that was common at the time. These bones have been designated as *Homo antecessor*, often known as Pioneer Man, with "The Boy of Gran Dolina" being the finest preserved. Despite its nickname, "The Boy of Gran Dolina", this kid was a female and was around ten years old when cannibals devoured her. The remains have garnered a great deal of interest due to the face's flatness. As we have seen, anthropologists and archaeologists seek for a flat face to identify whether an individual is a *Homo sapiens*. Compared to previous hominins, such as *Homo erectus,* and even later hominins such as *Homo heidelbergensis*, "The Boy of Gran Dolina" has a fairly flat face.

So, was this youngster one of our direct ancestors? Or was it one of our distant ancestors? Researchers determined that this was not the case. First, it is extraordinarily difficult to extract DNA from an 850,000-year-old fossil, but modern scientific approaches analyse specific proteins that survive better in the archaeological record. Proteins alone cannot provide as much information as DNA; therefore, please keep in mind that many assumptions are only interpretations of the data, which are subject to radical change in the future. Thus, proteins can provide information about ancestry, and it appears that *Homo antecessor* was possibly a sister group of our most recent common ancestor. A near relative who lived at the same time as us but was not our direct ancestor, the common ancestor, or an early member of the *Homo sapiens* lineage. The youngster of Gran Dolina exemplifies a complex dilemma faced by anthropologists. Even if a specimen shares many of the same physical characteristics as *Homo sapiens*, this does not always indicate that they are our direct descendants.

So far, neither the Boy of Gran Dolina nor any of the bones from the Sierra de Atapuerca in north-central Spain belong to our lineage. Given what we shall explain later regarding the genesis of our species in Africa, it is quite probable that our progenitor resided there. But because the past is so difficult to reconstruct, we must remain open to the possibility that our Neanderthal and Denisovan forebears existed in Eurasia. After all, that's where our two closest ancestors lived and evolved, so it's not unreasonable to assume that our common ancestor resided somewhere in West Asia, maybe even South Asia, such as India.

Wherever our ancestors resided prior to 300,000 years ago, it is quite obvious that Africa has been the driving force behind our evolution. This is supported not only by the fossil record but by human DNA as well. When attempting to identify whether or not a fossil represents *Homo sapiens*, anthropologists search for specific anatomical characteristics.

As we've already stated, we have a flat face that should rest right under our brain case, a very distinct chin, and a very spherical skull (globular is the correct name for this). Obviously, we must keep in mind that there is a great deal of variance in the appearance of modern humans, and that this variation goes far into the Paleolithic. So, these are not necessarily hard-and-fast criteria, but they are the characteristics that might suggest that a fossil is a *Homo sapiens*.

Dating back an astonishing 300,000 years, the earliest fossils that may represent *Homo sapiens* originate from Jebel Irhoud in Morocco, where five humans with all extraordinarily flat faces just beneath the braincase have been unearthed to date. Nonetheless, this skull is not spherical; it is more egg-shaped, like a Neanderthal, but more like *Homo heidelbergensis*. Due to the form of this skull, we cannot classify them as contemporary humans, yet they possess enough characteristics to be considered maybe ancient *Homo sapiens*. On the opposite side of Africa from Herto, Ethiopia, a skull that is around 160,000 years old was discovered in 1997. This skull is spherical but has extremely thick, chunky brow ridges, a highly ancient trait in our species.[20]

As seen by these two cases, humans in Africa during the Middle Paleolithic possessed a wide variety of features. Consequently, the development towards our contemporary form as anatomically modern humans appears to have been a relatively long process that spanned millions of years. This mosaic of characteristics has been preserved in the archaeological record for a very long period. The Omo remains are a collection of hominin bones found between 1967 and 1974 at Omo Kibish sites along the Omo River in Ethiopia's Omo National Park, with one of the relics being a very modern-looking, spherical skull. Research indicates that

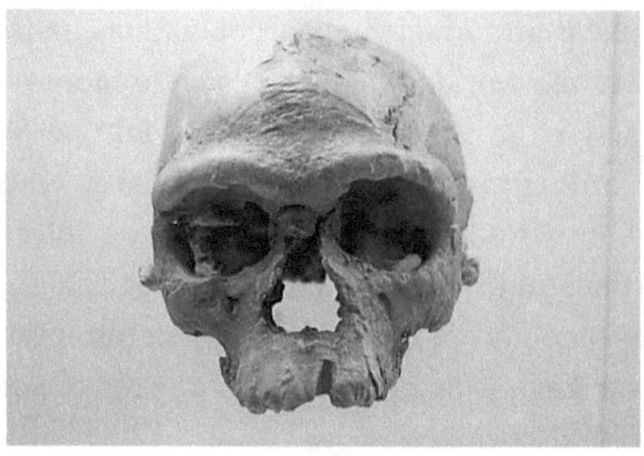

17) A reconstruction of Irhoud-1. A miner discovered a skull buried deep in the cave's wall, retrieved it, and handed it to an engineer, who kept it for a time as a souvenir. It was eventually given to archaeologist Émile Ennouchih, who subsequently unearthed another skull that he named Irhoud 1.

this Omo skull goes back 190,000 years. Another skull with a contemporary appearance was excavated in Qafzeh, Israel, dating back between 120,000 and 90,000 years ago.[21]

This convoluted picture of claimed *Homo sapiens* with diverse traits is made more difficult by the fact that we know other archaic human species were in Africa at the same period. *Homo naledi* existed between 236,000 and 335,000 years ago and appears to have been a substantially smaller species of hominin. It is unknown how these many archaic human populations interacted and what influence they had in the emergence of Sapiens. In the end, though, only we just remained.

Where have they all disappeared? What happened to them? Two competing hypotheses exist. The "Interbreeding Theory" gives a narrative of attraction, sex, and interbreeding that may explain what happened to our ancestors. The current human population is the result of interbreeding between

Homo sapiens and other human populations as they spread throughout the world. Neanderthals were there when *Homo sapiens* arrived in the Middle East and Europe for the first time. According to the Interbreeding Theory, when Sapiens travelled into Neanderthal territory, they interbred with Neanderthals until the two groups merged. If so, then the Eurasians of today are a hybrid of Sapiens (after the arrival of Sapiens in East Asia, they interbred with the indigenous Erectus).

The opposing hypothesis, known as the "Replacement Theory," paints an entirely different tale: incompatibility, repugnance, and potential genocide. *Homo sapiens* and other humans had diverse anatomy and maybe even separate mating practices and body odours. They would not have shared any sexual attraction, and even if a Neanderthal and a *Homo sapiens* attempted to mate, it would be impossible to produce fertile offspring because of the enormous genetic divide between the two groups. Thus, when Neanderthals and other human species went extinct or vanished, their DNA died with them. According to this view, *Homo sapiens* replaced all other human species without interbreeding with them. If this is true, then all *Homo sapiens* on the earth today may be traced back to East Africa roughly 70,000 years ago.

This point is essential since 70,000 years is a very short period of time in terms of evolution. If the Replacement Theory proves to be accurate, all humans alive right now have roughly identical genetic backgrounds, and racial variations are negligible. If the Interbreeding Theory is correct, however, genetic differences between Africans, Europeans, and Asians may extend back hundreds of thousands of years!

In recent decades, the Replacement Theory has gained widespread acceptance in its field due to its increased archaeological support and more political acceptability. In 2010, the results of a four-year effort to map the Neanderthal genome were published, putting an end to this debate. Geneticists retrieved sufficient intact Neanderthal DNA from remains to conduct a thorough comparison with human DNA. Scientists were stunned to discover that between 1% and 4% of the unique human DNA of modern people in the Middle East and Europe contained Neanderthal DNA. Obviously, this is not a big amount, but it is considerable. Several months later, the DNA extracted from Denisova's fossilised finger was mapped, demonstrating that up to 6% of the unique human DNA of contemporary Melanesians and Aboriginal Australians contains Denisovan DNA.

However, it is important to remember that further research is currently being undertaken, which may either confirm or alter these results. The biological cosmos is not binary, as there are also substantial grey areas. There must have been a moment when the two groups were so unlike that they could only occasionally produce fertile offspring. Then, another mutation destroyed this final link, forcing their evolutionary trajectories to separate. Approximately 50,000 years ago, it appears that Sapiens, Neanderthals, and Denisovans were on the edge of this thin thread. They were virtually different species, but not quite.

But, returning to the Omo skull in Ethiopia, is it a late survival of one of these ancient human forms? Or have Sapiens in this location kept archaic characteristics for a longer period of time than anywhere else? In the archaeological record, there is a distinct shift from egghead to globular-shaped skulls. But what areas of our brain changed and developed to generate this shift in form? And, most importantly, what does this tell us about our evolution?

Comparing two brains, two endocasts from our evolutionary past (one from 300,000 years ago, excavated in Jebel Irhuod, Morocco, and a modern human brain), both brains have nearly the same capacity and size, but they have been reorganised and reshaped.[22] So, what parts grew and evolved, and what conclusions can we draw from that?

Two characteristics of this process include perinatal and cerebellar enlargement. Orientation, attention, stimulus perception, sensory-motor transformations, underlying planning, physio-spatial integration, imaging, self-awareness, working and long-term memory, numerical processing, and tool usage all include parietal regions. In addition to motor functions such as movement coordination and balance, the cerebellum is connected with spatial processing, working memory, language, social cognition, and effective processing.

Increasing our inventiveness, perceptions, and flexibility is both a result of evolutionary pressure and an evolutionary benefit. Given the abundance of creativity that surrounds us daily, this may seem clear. Nonetheless, it is intriguing to consider this brain development. This transition from an egg-shaped to a sphere-shaped skull during the past 300,000 years is represented in the archaeological record by an increase in creativity as evidenced by a rise in the sophistication of art and tools.

In South Africa, ocher-carved ostrich eggs, and shell-crafted jewellery serve as artistic and symbolic expressions of the highest calibre. Approximately 50,000–30,000 years in the past, this spherical brain form is nearly universal in *Homo sapiens*, with a few exceptions, and the evidence for symbolic communication and specialised equipment is overwhelming. No one could look at this period's artwork and conclude that these people were simply cavemen. Nothing could be farther from the truth since no archaic human has ever demonstrated the same level of ingenuity and symbolism exhibited by contemporary humans.

The genetic evidence for our evolution in Africa is overwhelming and nearly indisputable. African populations have higher genetic diversity than populations anyplace else in the globe, and genetic diversity declines as one moves away from Africa. This demonstrates that humans almost certainly evolved in Africa.

This is known in biology as the Stichtereffect (Founder effect). If you have a location where a species evolved, you will find the full genetic variation of that species in that region. However, if a group splits and relocates to a different zone, the new region can only support the diversity of the group that split. This tendency is evident in the genetic record (high diversity in Africa, lower diversity outside Africa).

Consider the mitochondrial haplogroup L3. Haplogroup L3 is closely associated with the "out-of-Africa migration" of modern humans between 75,000 and 50,000 years ago. It is inherited by all non-African contemporary groups, as well as by certain African people. A haplogroup is a collection of alleles (a specific gene variation) on various chromosomal locations that are tightly connected and typically inherited together. Essentially, it is a lineage of ancestry that can be traced back to a single individual in deep prehistory.[23] (For simplicity, let's call "mitochondrial haplogroup L3" just "Group L3").

Group L3 is, therefore, the modern-day descendant of an ancient L3 lady who lived in sub-Saharan Africa. This group is particularly fascinating because the haplogroup from which every ancient fossil for which we have DNA originates outside of Africa. In addition, every non-African haplotype derives from this haplogroup, indicating that all mitochondrial haplogroups outside of Africa are descended from the L3 Group. Therefore, if you are not African and you are reading this, you are nearly definitely descended from this L3 woman who was African. No ifs, buts, or maybes exist. No matter where you're from (from Europe, Asia, Australia to the Americas), you descend from this L3 woman. Even if you are from Africa, there is a strong possibility that you are a descendant of this woman.[23]

Now mitochondrial DNA is a very small portion of our genetic composition, and this has been repeatedly validated. In 2016, research integrated hundreds of DNA testing to generate a genetic map of humankind. The same pattern strongly suggests that Africa is the origin of our DNA. This pattern may be observed not only in our own genetics, but also in the genetics of the microorganisms that reside within us. *Helicobacter pylori* is a gut-dwelling bacteria that cause stomach cancer and ulcers. This bacteria is very interesting because we observe the same pattern of genetic variety: Helicobacter in Africa has a high level of genetic diversity, but outside of Africa, it has a low level of genetic diversity.[24] This suggests that this bacterium was present in *Homo sapiens* that first left Africa and suffered a loss of genetic diversity as a result, much like *Homo sapiens*.

As it stands, the oldest *Homo sapiens* fossils, or archaic *Homo sapiens* fossils, comes from Africa, and genetic evidence strongly implies that Sapiens evolved in Africa. However, this region is vast. Can we build upon it? Is there a particular location in Africa where humans evolved?

If we examine the fossils we have already stated, such as those from Herto, Omo, and Qafzeh, we see that these purported *Homo sapiens* had a vast pan-African distribution. These are the legitimate candidates for ancient Sapiens (however, not every scholar agrees on their classification). The only location where there appears to be a divide is in West Africa, but this region has been the least investigated archaeologically, so it's possible that this is merely a bias in the archaeological record.

Regardless, there is a wide spread of these ancient *Homo sapiens* from north to south. Given this dispersion, it might be challenging to pinpoint a certain time when Sapiens evolved. There is no difference in the genetic record: various research has reached divergent results as to whether South, East, West, or Central Africa was the continent's initial location.

Isn't there substantial proof that modern humans are descended from a big population? And how can a single woman and a single man be the forefathers of everyone? The simple answer is that a single man and a single woman are our shared ancestors, but they are not our sole or unique forebears. Instead, they are both members of the "Mitochondrial Eve and Y-chromosome Adam" population group. These imaginary individuals existed around 200,000 years ago. Suppose we discovered the location in which they resided: that is then not certainly the cradle of humankind, but these two individuals are part of a group whose genes have likely persisted to the present day. In conclusion, even if anthropologists and archaeologists are quite sure that *Homo sapiens* evolved in Africa, what we mean by this and precisely how this evolution occurred are still highly abstract and hotly debated. However, it poses major questions if humans evolved in Africa. Why did we leave? And most significantly, when did we leave?

If you are not African, your ancestors left the continent some 60,000 years ago. This estimate is not based on archaeological data but rather on genetics. Do you recall mitochondrial haplogroup L3? This branch of our family tree, this lineage, was established around 60,000 years ago and completed our migration out of Africa. Our forefathers did not leave the continent until around 79,000 years ago (other genetic research have placed the migration out of Africa at around the same period, about 55,000 years ago). What about the Helicobacter? Their genetic split happened around 58,000 years ago (Thus relatively around the same period as that of *Homo sapiens*). Consequently, many genetic tests and evidence imply that our ancestors departed Africa between 55,000 and 70,000 years ago.

Out of Africa

However, there is one significant problem: Archaeologically speaking, we seem to have evidence of *Homo sapiens* outside of Africa much earlier than that, which we call "Out of Africa I."[25] As we have seen, *Homo sapiens* lived in the Levant between 90,000 and 120,000 years ago, and recent research published in 2021 demonstrates that there have been several migrations into western Asia during the previous 400,000 years. This location is adjacent to Africa and may be reached with a quick hop, skip, and jump. In the north, it is joined to the continent, and in the south, the historic strait between Asia and Africa was just 4–15 kilometres wide. Asia was seen just beyond the shore.[8]

Western Asia is not the only place where very ancient *Homo sapiens* bones have been discovered. In Greece's Apidima cave, archaeologists discovered a *Homo sapiens* skull dating to around 230,000 years ago. In addition, *Homo sapiens* teeth dating between 80,000 and 113,000 years old have been discovered in China. Around 70,000 years ago, *Homo sapiens* were in Sumatra, Indonesia, and by 65,000 years ago, we were on the northern coast of Australia. These final two are inside the timeframe when humans would have left Africa, as suggested by our DNA records and research, although they are quite close to the window's edge. We would have had to race around the south coast of Asia to get there in time, which is plausible, but we still have those remains in Greece, those remains in China, and a large number of hominin tools in India that we haven't been able to attribute to a particular individual.

How can we thus reconcile the conflicting DNA and archaeological evidence? The first option is that we overlooked genetic cues from the initial migratory wave. In 2016, geneticists revealed the discovery of a previously uncovered genetic signal in the genomes of Papua New Guineans, indicating that they contributed to their *Homo sapiens* DNA about 120,000 years ago. It was a small amount of their DNA: about 2% (thus, they are still primarily cousins of the later African migration). Nevertheless, it is plausible that we overlooked these genetic markers from an earlier migration.[25]

However, other researchers that knew geneticists and had made this conclusion have attempted to duplicate it without success. Consequently, it is currently up for dispute. It is yet unproven that this genetic signature represents a historical migration from Africa. The alternative argument is that these previous migrations of *Homo sapiens* from Africa left no genetic record, indicating that they were not indeed our direct ancestors and eventually failed. This raises a major question: Why did ancient humans such as Neanderthals and Denisovans survive, whereas the first migration wave of *Homo sapiens* became extinct?

But if these "first-migration-wave" *Homo sapiens* societies lasted until this later migration, why do we not find more evidence of interbreeding between our two lineages? Especially in light of the abundance of evidence for interbreeding with other ancient humans? The research of ancient DNA in Europe and early European fossils has revealed that we are not closely connected to the original Europeans and that there was a significant population decline. Consequently, it is probable that these previous migrations have ceased to exist.[26]

Whatever made the second expansion wave from about 55,000–70,000 years ago (known in this context as "Out of Africa II") of *Homo sapiens* successful, it is clear that we absolutely stepped on the scene.[25] We kicked in the door and rampaged across the world like an invasive species. *Homo sapiens* reached Europe and the far east coast of Eurasia around 50,000 years ago. Around that time, we travelled

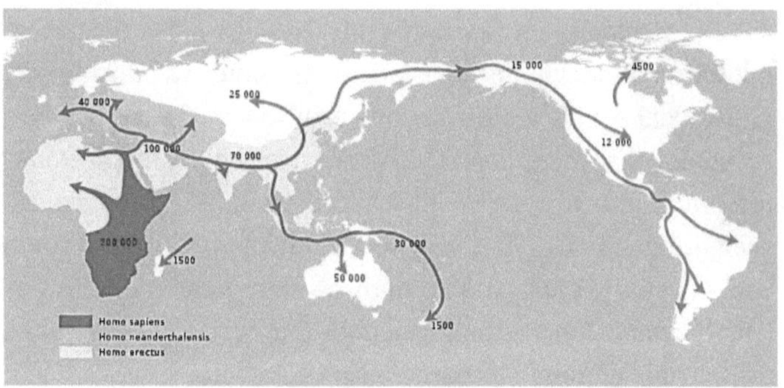

18) The migration map of modern humans, including *Homo sapiens*, *Homo neanderthalensis*, and *Homo erectus*.

through Indonesia's northern islands, into New Guinea and eventually entered Australia approximately 45,000 years ago.

And finally, *Homo sapiens* entered North America around 25,000 years ago via the Bering Strait land bridge, which had formed between northeastern Siberia (Russia) and western Alaska (America) due to the lowering of sea level during the Last Glacial Maximum (the last Ice Age). Those humans that made it across the Beringia landbridge and spread out into North, Central and eventually South America have only a few parents of their genetic origin. Professor Jody Hey suggests that the first original population of the Americas was around eight families strong (about 70 humans). However, eventually, climate change at the end of the last Ice Age caused the glaciers to melt, flooding Beringia about 10,000 to 11,000 years ago and closing the land bridge, making the crossing between Asia and the Americas impossible. It was then that they became estranged from one another, and for almost 10,000 years, they would go their own ways, never again meeting or even knowing of one another.

Arctodus—an extinct genus of short-faced bears that inhabited modern-day Alaska, North America, during the Pleistocene—was a furious, powerful beast. There are two recognised species: the lesser short-faced bear (*Arctodus pristinus*) and the giant short-faced bear (*Arctodus simus*). Weighting up to 950 kilograms, equivalent to two brown bears (*Ursus arctos*), and up to three metres tall on their rear legs, *Arctodus simus* is believed to be one of the largest known terrestrial mammalian carnivorans that have ever existed! Both short-faced bear species are relatively rare in the fossil record, but many suggest that these species (especially the giant *Arctodus simus*) were one among the numerous factors that kept humans from crossing this Bering Strait.

It's unlikely that the first adventurous explorers to cross this bridge and catch sight of this bear had any intention of using the bear's fur to fashion a cosy blanket (fleeing would have been a more likely and wiser option). However, brave humans eventually successfully crossed this land bridge, marking the beginning of *Homo sapiens* footprints in the Americas. Everywhere we went, populations of large animals seemed to die out shortly after. We went all out, and there's no doubt about it: *Homo sapiens* is an ecological serial killer.

But what gave us that killer edge in the second expansion wave? What made us so much more successful than any other archaic humans and the previous *Homo sapiens* migratory wave that preceded us? In addition to what we have lately mentioned, such as intricately interwoven fiction, complicated language, communication, and our refined cultures, it is likely that our advanced technology would have been the key to our success.[27] In warfare, technological

19) A reconstruction of a giant short-faced bear (statue) compared to a *Homo sapiens*. A mature short-faced bear could reach speeds of 60 kilometres per hour, making successfully fleeing difficult. It was quite a scary predator of its time, wouldn't you say?

innovation is incredibly important. Gaius Julius Caesar and his brave army would have little chance against today's drones and guided missiles, regardless of how magnificent, strong and brave they were.

About 48,000 years ago, a group of Sapiens named the deep jungle in what is now southern Sri Lanka "home." The rainforest is actually a very different ecosystem from the African savanna, where humanity most likely evolved. However, although the rainforest is extremely verdant and lush, it is not an easy place to live in. Nevertheless, this prehistoric group of Sapiens successfully adjusted and left behind a treasure trove of arrows crafted from the bones of monkeys, as discovered by archaeologists. One of the arrows is even etched with lines that other hunter-gatherers used historically to attach poison to arrows.

As it is difficult to hunt the tiny and nimble animals that inhabit the thickly forested area, technology such as the bow and arrow must have been a significant component of their survival strategy. Intriguingly, the oldest archaeological evidence of bows and arrows comes from Africa between 60,000 and 70,000 years ago, at the beginning of our journey. It's fascinating to discover that this rainforest hunter-gatherer group was contemporaneous with the inhabitants of Siberia's far north.

The progenitors of these two populations of *Homo sapiens* may have departed Africa barely 15,000 years ago. One group migrated to the north and acclimated to the cold, while the other embraced the south coast of Asia and adapted to the rainforest. This degree of technological adaptability and sophistication is exclusive to modern humans, not to other archaic humans, archaic *Homo sapiens*, Neanderthals, or Denisovans. As the climate around them changed, these contemporary humans adapted by following their chosen habitat. This latter migration of *Homo sapiens* was able to adapt by modifying their lives and adopting novel, inventive technology to survive more efficiently in their environments (exactly as Charles Darwin said in *On the Origin of Species*: "survival of the fittest"—organisms that are most adapted to their environment are the most effective at surviving and reproducing).

We, *Homo sapiens*, have spread all over the planet. After 30,000 years ago, all our archaic cousins seem to be gone. But although we have been on this intense, fascinating journey for millions of years, this point is actually just the very beginning of modern human history (what we consider the Neolithic revolution, also known as the Agricultural

Revolution). This chapter ends at the point where modern human history begins. However, as we shall see in the book's next chapters: this isn't the end, it's not even the beginning of the end, but it's the end of the beginning.

PART III

THE IMPERIAL

REVOLUTION

0) The Colosseum, constructed in 72 CE and is one of the Seven Wonders of the World, exemplifying the refined aesthetic culture of ancient Rome. The Colosseum is a symbol of ancient Rome and is regarded by many as the zenith of ancient civilisations.

The Cradle and Rise of Civilisation

8. THE AGRICULTURAL REVOLUTION

About 70,000 years ago, the Cognitive Revolution gave *Homo sapiens* the ability to converse about things that were only in their own minds. *Homo sapiens* weaved several fictitious webs during the ensuing 60,000 years, but they remained modest and regional. Beginning around 35,000 BCE, modern humans' ability to imagine great spirits and create new tools blossomed, as evidenced by an increase in the number and sophistication of bone flutes, figurines, and other works of art.

Yet, one tribe's beloved ancestral spirit was unknown to its neighbours, and shells that were precious in one location lost their value as soon as you crossed the surrounding mountain range. Stories about ancestors' spirits and priceless shells continued to provide humans with a significant advantage because they allowed hundreds, sometimes even thousands *Homo sapiens* to cooperate in a way that neither chimps nor Neanderthals could. Because it was challenging and unsustainable to feed a city or a kingdom by hunting and gathering, Sapiens were unable to collaborate on fairly huge scales as long as they continued to be hunter-gatherers. As Harari states, "The spirits, fairies, and demons of the Stone Age were, therefore, rather weak beings".[0]

For millions of years, humans sustained themselves by collecting vegetables and hunting wildlife that lived and mated without their interference. *Homo erectus* and the Neanderthals foraged at wild figs and plants and hunted wildlife without considering where plants and fig trees would take root or where a herd of sheep would graze. Even as *Homo sapiens* moved from East Africa to the Middle East, Europe, Asia, and, finally, Australia and the Americas, they continued to hunt and gather. Of course, it was natural for these human species to stick to this lifestyle: why do anything else when your lifestyle feeds and sustains an exceptionally diverse and sophisticated world of social systems, hierarchy, religious movements, and political structures?

The Last Ice Age

Beginning around 2.4 million years earlier, the last Ice Age's colossal ice sheets that covered most of Asia, Europe, and North America ceased their slow march some 27,000 years ago. *Homo sapiens*, despite their difficult circumstances, were living rich lives full of ritual, symbolism and creativity. However, the threats of their world were always there.

Approximately 27,000 years ago, the last Ice Age reached its peak extent, with large glaciers covering entire continents (North America and Europe were covered by an ice sheet approximately three to four kilometres thick). What land isn't buried under glaciers has become a mixture of forests, steppe and tundra. Woolly rhinos, cave lions destined to become the Ice Age's apex predator, cold-adapted hyenas, and plenty of other modern-day species all called this region home. However, the mammoths (*Mammuthus primigenius*) were the true awe-inspiring giants of this Ice Age grassland, the rightful apex predators of the steppe.

In yet another Stefan Milosavljevich documentary,[18] a young man known as "il principe," also known as "the prince",

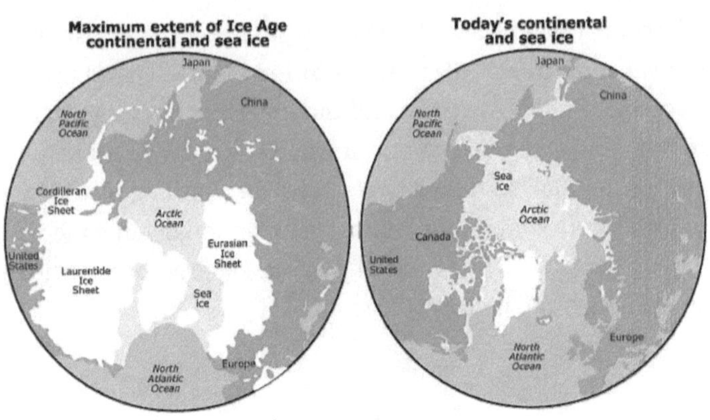

1) A map depicting the greatest extent of the ice sheets during the last Ice Age and their present extent.

who lived during the peak of the last Ice Age, in a cave overlooking northern Italy and the Mediterranean Sea, is described. Nicknamed because of his lavish burial with hundreds of shells, an exotic flint blade, some long-lost medicine, huge ornaments, and unfathomable tools, we're going to reconstruct as far as possible the life and death of this palaeolithic prince and the lives of people who lived around 27,000 years ago in what archaeologists call the Gravettian period (an archaeological industry of the European Upper Paleolithic art and culture).[15]

Probably no older than 16 or 17 years old at the time of his death, il principe was just a young man when he died. Cooperation, social complexity, and in-depth technological understanding are crucial to our existence, especially amid the harsh conditions of the Ice Age. It's impossible to be born with those abilities, to be able to talk and create tools. These are things that can only be taught with time, during childhood, and during a lifetime.

Finding useable evidence of children at ice period archaeological sites can be challenging but not impossible. Parents always loved their kids, and that hasn't changed. Children were occasionally buried, as evidenced by sites like Sungir and Krems-Wachtberg, and they were often buried with objects that modern-day kids would be familiar with.

However, they also left behind more than simply toys from the prehistoric era. They left behind traces of their existence in the form of cave art, one of the most fascinating remnants from the Ice Age. The handprints our ancient forebears left in caves all around the world are, without a doubt, the most impressive and recognisable artefact we have from their time. Africa, the Iberian Peninsula, France, Indonesia, Australia, and even southern Argentina have plenty of cave art artefacts. They make such a classy and lovely proclamation: "I was here. I existed." It's innately human to want to make a lasting impression on the world. Children were present throughout the creation of this artwork, and some of them contributed to it. Archaeologists can infer the artist's age from the size of their handprints: up to 27% of the hand drawings in caves like Fuente de Salin on the northern coast of Spain were likely made by children! Some of these small handprints are too high for a child to have made on their own, so a grownup, perhaps one of their parents, must have raised them. Artists worked side by side, creating art that held such symbolic and intriguing meanings for both of them.

A little necklace made from mammoth ivory hung around Il principe's left wrist as he was laid to rest. We can only speculate as to its symbolic importance to him, although others have suggested that it is a stylised version of a female

2) Hand prints in Pettakere Cave at Leang-Leang Prehistoric Site, Maros. Scientific examinations conducted in 2011 estimated that the hand stencils on the walls were between 35,000 and 40,000 years old.

figure. No of the interpretation, it is unmistakable evidence of mammoth hunting, and it might have been a souvenir from a hunt in which he took part. It's safe to assume that hunting Gravettians were well-versed in the use of flint tools. The ability to do this is crucial for success in their lifestyle, so it's hardly shocking that they would bury a family member with a valued possession like a beautiful blade.

These giants of the steppe would have been difficult opponents for il principe and his band, standing over three metres tall, weighing up to six tonnes (about the weight of two pickup trucks), and having tusks that might grow to a length of four metres. Hunting them was a tempting option, with the potential to feed entire communities, supply a massive amount of bone and ivory for tools, a massive amount of fur, and maybe even shelter, but also the possibility of getting gravely injured or perhaps killed.

The archaeological record is rich with clues as to how our Ice Age ancestors went about hunting these creatures. There is one mammoth in particular that provides some striking insights. A mutilated mammoth was discovered at the northernmost tip of Siberia. At about 45,000 years old, it is significantly older than the prince. It is the oldest direct evidence of humans in the Arctic, and thanks to its remarkable preservation, we can get a fantastic understanding of the methods used to hunt these majestic creatures.

In order to kill a mammoth, the hunters have to encircle it. Several of them launched spears at the mammoth's chest, smashing bones on the way to the mammoth's internal organs and tearing the ribcage. Bleeding and terrified, the mammoth becomes weaker. To prove their bravery, one hunter approaches closely and thrusts his spear into the victim's shoulder. After the six tonnes mammoth collapses, the hunters approach the elegant animal. After avoiding the mammoth's massive tusks, the brave hunter strikes his spear downward into the mammoth's cheek, perhaps aiming for the nasal passage, which was the killing blow to end his agony. (Traditional African elephant hunters generally aimed for the base of the trunk, where the animal's primary arteries are located, to ensure a quick and easy kill. The giant animal would have had to be already down for this strike to be possible).[17]

The prince of the time had no choice but to hunt for his survival, and while the sight of blood and gore may be unsettling to us in our world today, we must keep in mind that the Ice Age was the most dangerous time for *Homo sapiens* in history! Warmth, shelter, and plenty to eat were all made possible by the mammoth's death. Evidenced by their

3) A representation of how il principe and his band hunted these majestic Ice Age beasts.

depictions of ancient animals in art, the inhabitants of this era clearly felt a strong connection to the wild creatures that shared their world. The killings weren't mindless butchery at all. During the Ice Age, humans relied on the hunting of mammoths.

The shallow Beva valley in modern-day Moravia, eastern Czechia, was apparently also ideal for a community of mammoth hunters. Over a thousand mammoth remains were discovered in a location known as Predmosti. Additionally, 20 human tombs were also discovered, and from them we learn that the Predmosti hunters had access to a new, especially "fluffy" technological advancement that aided them in their hunt for mammoths.

At the site, archaeologists uncovered more than 4,000 canine bones. Many of these belong to what appear to be wolves. Intesteringly, three of the skulls were significantly smaller than is typical for wolves. In fact, they are incongruous with any known wolf skull. In all likelihood, these canines date back to prehistoric times. It's possible these were hybrids between a wolf and a dog. Their connection to Predmosti's hunting community is unambiguous. These weren't simply lounging outside the camp, as some were discovered in graves with humans, others had their teeth artificially altered, and one even had a mammoth tusk inserted into its mouth. Based on the available data, it's safe to say that this was a group of mammoth hunters who were also active in wolf domestication. Given the kind and age of the mammoths discovered there, it's possible that dogs and wolves assisted the hunters in isolating the young of the herd and pursuing them for kilometres along the valley's edge, tiring them out enough that they could be killed with relative ease.

According to the analysis of grinding implements discovered in Italy from this time period, besides the pursuit of meaty-rich meals, cattail and acorn roots were ground into flour. A plain loaf baked in the ashes of the campfire would have been perfect for the il prince if they were milling flour. The prince's daily food was likely far more modest than the hunting of mammoths, which is, of course, spectacular and exactly what we imagine when we think of the Ice Age. He sat around the fire with his loved ones, talking about the day's catch, the nuts and seeds he had gathered, the small game they had hunted, the roots they had dug up, and the bread they had baked.

Some very peculiar handprints date to the Gravettian period and have been discovered in caves in contemporary France and Spain. Fingers and joints are clearly missing from what seems to be a series of handprints. Of the 231 drawings of hands found in France's Gargas cave, 114 are missing at least one finger. Of the 49 depictions of hands found at Cosquer cave, 28 are incomplete. In Maltravieso, in the west of Spain, 61 of the 71 paintings of hands do not have all their fingers (There are about 200 total examples, all dating back to the Gravettian era in France and Spain). Considering the frigid climate of the time, it's possible that some people lost fingers to frostbite, but the prevalence of depictions depicting hands with amputated digits raises the possibility that this was done on purpose. This may seem like a bit of an exaggerated claim by some archaeologists, but it is actually a well-documented phenomenon.

One research identified 121 different societies from all across the world that practised the intentional amputation of fingers. Among their daily rituals, the Dani people of Indonesia cut off their fingers after the death of a loved one. If the Blackfoot of North America needed the Sun god to save or defend a life and a huge sacrifice was necessary, someone's finger would be severed as part of the Sundance ritual. In southeast Africa, among the Ila-speaking people, amputation of fingers was a common punishment for offences including adultery and stealing. Daughters of aboriginal families in the Port Stephens area of Australia would have two joints cut off of their fifth fingers if they wanted to be identified as fishermen. Consequently, there are numerous examples of this practice, carried out for a wide range of purposes, in the anthropological record from all across the world.

Despite the fact that il principe had all his fingers, he was fatally wounded when he was still young. Il Principe was apparently knocked unconscious by a blow to the face, as evidenced by the missing bone from the left side of his jaw. His unexpected death may have been the consequence of a hunting accident in which he was gored in the face by an Ice Age goliath. Maybe he was the victim of a violent attack by someone else who was angry with him over anything (love, envy, resources, or power). The prince's wound had obviously started to heal, but it was too late.

Even though he was buried like a prince and was known as il principe, it doesn't mean he was treated like one in his hometown (no data exists to back up such a claim). In addition to being utilised as votive offerings to the gods, weapons (such as the prince's blade) were also handed to the families of fallen soldiers as part of their burial rites. The same was true of metal smiths. Divine intervention was sought for both the smelting of ore into a usable artefact and the actual combat and war that ensued. As a result, smithing tools are often included in votive offerings. One practical purpose of these sacrifices was to display the sacrificer's riches and power. Similar considerations should be given to grave goods, as they allow the deceased's family or tribe to demonstrate social status through elaborate funeral rites. Communities would often gather at the tombs of the elite to reflect and pay their respects. The veneration of fictitious forebears was used as a political tool to solidify the present ruler's position.

Yet, whatever it means in your culture, most people just want to say goodbye properly to their loved ones. The fact that he was honoured with such a lavish burial is no guarantee of his social standing (he may just have been very liked). He was laid to rest alongside the precious artefacts of his period, which have been with him for over 27,000 years and have given us a glimpse into the life of this prehistoric Ice Age human.

The Great Flood

Approximately 25,000 years ago, roughly 2,000 after the death of il principe's death, the Ice Age glaciers mostly retreated and melted away as the planet warmed after the Pleistocene ended. Yet, some ice cover has stood the test of time, such as the glaciers in the Antarctic Peninsula. The thawing of the glaciers was a key turning point in the history of *Homo sapiens* since it not only transformed the status of humankind but also paved the way for civilisation. However, before we can move on to the birth and evolution of civilisation, we must first ask and answer a very important mystifying question: What caused the abrupt end of the Ice Age?

There are several hypotheses on what caused the end of the last Ice Age, the most scientific explanation of which is from the latest research from the University of Melbourne, which states that the end of the last million years' ice ages happened when the tilt angle of the Earth's axis hit higher levels. During these epochs, longer and stronger summers melted the enormous ice sheets of the Northern Hemisphere, causing the Earth to enter a warm "interglacial" phase like the one we have been experiencing for the past 11,000 years. Yet, researchers are still striving to identify the frequency and timing of these occurrences.

However, the Younger Dryas Impact Hypothesis is a more controversial alternative theory. According to this hypothesis, between 13,000 and 9,000 years ago, one or two fragments of a comet crashed with Earth, causing a dramatic climate shift and the widespread extinction of Ice Age megafauna. At the end of the last Ice Age, the rapid rise in sea levels was driven by the melting of the world's ice sheets. It is thought that this devastating catastrophe inundated a region greater than both China and Europe in a single, apparently instantaneous flood.

The legendary ancient Greek philosopher Plato estimates in his book *Laws* that this flood occurred roughly 10,000 years before his time, and the latest geological study presents the intriguing possibility that Plato's tale is not entirely fictional. Astonishingly, the year 9600 BCE that he provides aligns with an extremely accelerated sea level rise, also known as the Meltwater Pulse 1B disaster (this may have been the catalysis of the Atlantis story). Although many of these stories offer explanations for the catastrophe, they all share the notion that a great flood swept out an ancient civilisation for its sins and ushered in a new era of human history. While these stories may not be founded on solid evidence, they are still significant in light of the increased risk of floods generated by glacial meltwater and the pervasive threat of natural disasters in the modern world.

Journalist Graham Hancock claims that there is no need to debunk myths: "they are the keepers of our species' collective knowledge; thus, we should always listen to what they have to say." Atlantis is not an unconnected story: Hancock argues that hundreds of oral tales from all around the world describe a catastrophic flood that swept away an advanced Ice Age civilisation and terminated an age of unprecedented prosperity.

Since the world seemed drastically different during the last Ice Age, it is intriguing to ponder what may be lost. With one of the biggest known river networks up until the end of the Ice Age, flora flourished in the Sarah Desert's rich soil, giving the desert a remarkably lush appearance. However, because it is so hostile nowadays, modern archaeologists haven't made many efforts and discoveries there. In addition to the Sarah Desert, archaeological sites have been discovered in the Amazon rainforest, where LiDAR scans (light detection and ranging) have shown gigantic constructions hidden beneath the canopy.

The Spanish coloniser and explorer Francisco de Orellana accidentally crossed the Amazon River in the 1560s. Orellana and a group of 20 men headed off on a hunting excursion, but the stream they were following wouldn't let them turn around, so it carried them hundreds of kilometres west until they reached the Atlantic coast of South America. During their detour, they came across large cities and were stunned by their size. At the time, his claim was received with scepticism, but as just stated, LiDAR scans have confirmed that he was correct.

However, returning to the controversial, yet interesting statements of journalist Graham Hancock, the fact that many missing Ice Age puzzle pieces are slowly coming to light makes his claims tough to debunk. In 1513 CE, the Ottoman admiral and geographer Piri Reis created a world map that is now known as the Piri Reis map (the map was compiled from a large number of ancient maps, among them Christopher Columbus, four Portuguese maps, an Arab map of India, eight maps drawn during the time of Alexander the Great in 332 BCE and six maps that were from unknown sources).

This map is historically important because it shows how far Europeans had already travelled in the New World by 1510 CE. Only one-third of the original map has been preserved, but the remaining bit accurately shows the western shores of Europe, North Africa, and Brazil. Several Atlantic Ocean islands are depicted, including the Azores, the Canary Islands, and even the mythological Antillia. Beginning around Rio de Janeiro, this latter path is portrayed as abruptly turning east. The idea that this area is identical to the shore of Queen Maud's Land in Antarctica has also sparked conspiracy theories.

Hancock thinks it intriguing that Antarctica is shown on the Piri Reis map, given that the continent was not discovered until the First Russian Antarctic Expedition in 1820 CE. Hancock believes that we are dealing with the fingerprints of a lost civilisation that mapped the planet and left evidence that ancient mapmakers discovered, utilised, and integrated into their own maps. In spite of the increased attention devoted to these controversial hypotheses in recent years, academics have yet to find a convincing explanation.

The end of the Ice Age, on the other hand, signified the beginning of civilisation. Regardless of what caused the end of the Ice Age or what evidence of it is still buried, we can most confidently state that after the end of the Ice Age, the cradle of civilisation began to mature, eventually flourishing. However, this groundbreaking development was not without problems and consequences for *Homo sapiens*.

The Dawn of Civilisation

Around 12,000 years ago, with the melting ice and the warming of the Earth, at the start of the Neolithic, or New Stone Age, humans embraced a more sedentary lifestyle and began to dedicate practically all of their time, effort, and freedom to manipulating the lives of certain animal and plant species. However, with the end of the last Ice Age paving the way for a different lifestyle, this sudden change in diet and lifestyle had a serious impact on human health. Sapiens planted and sowed seeds under the blazing Sun from dawn to sunset, watered plants, plucked weeds from the land, and guided sheep to optimised fields. These forward-thinking Sapiens believed that this would provide them with adequate grain, meat, and fruit. It was indeed an important revolution and would forever change the entire course of *Homo sapiens*: the Agricultural Revolution.[0]

Around 9500-8500 BCE, the Agricultural Revolution started in the southeastern section of Turkey, Syria, western Iran, and the Levant. However, it began slowly and in a limited geographical region, and archaeologists disagree over how it unfolded. While some archaeologists believe agriculture originated in a single location in the eastern Mediterranean and expanded from there, others believe it had numerous independent origins, a theory that is gaining traction.

During this Agricultural Revolution, humanity's technological capabilities exploded. Among the wheel, stone tools, and the plough, pottery is widely regarded as one of humanity's greatest inventions from this historical era. Much of pottery's rich history may be traced back to ancient, pre-literate cultures. Because of this, archaeology is essential for uncovering much of this past. Pottery and shards of pottery are among the most common and significant artefacts found at archaeological sites because of their longevity over the millennia.

Fired clay pieces, known as ceramics, are among the most valuable artefacts that can be uncovered at excavation sites. Ceramic artefacts are extremely durable and may last thousands of years virtually unchanged from the date of manufacture. Furthermore, unlike stone tools, ceramic artefacts are entirely crafted by humans, taking the form of clay before being kiln-fired for decorative or functional purposes. By the Neolithic Age, pottery made from modelled clay had become an independent art form and was utilised in performing religious rituals, preparing food, and transporting liquids and solids.

It's possible that pottery was accidentally crafted in the ashes of different fires on different clay soils, leading to their discovery independently in various places. Pit firing and coiling were the two first methods of making vessels, and both are quite easy to master. The Venus of Dolní Věstonice figurine, excavated in the Czech Republic, is made of ceramic and dates back to 29,000–25,000 BCE, making pottery one of the oldest human inventions.

And yet, Eastern Asia is widely recognised as a key region for the development of "functional" pottery. This is especially true regarding pottery vessels—used for storing, cooking and serving food and carrying water—that were discovered in Jiangxi, China, which date back to 18,000 BCE. Throughout the Mediterranean, pottery emerged as a separate art form following the introduction of agriculture. Farming settlements emerged in the Levant region of the Near East some 12,000 years ago. However, they didn't develop pottery for another 3,000 years (we called the time between these two dates pre-pottery Neolithic). They were very inventive when it came to food storage. Excavations in Jordan revealed granaries that even had raised floors to keep some vermin out to allow some air to circulate from as early as 11,000 years, and they probably produced alcohol in stone troughs several thousand years later.

Today, alcohol is a drug, but for some of our earliest evolutionary ancestors millions of years ago, the smell of fermentation was a signal that fruit was at its ripest and most calorically dense. The discovery of how to ferment and distil alcohol using pottery, however, had great far-reaching implications for early human societies. First, they had important nutritional value; Second, they were the best medicine available for some illnesses and especially for relieving pain. Also, those who drank fermented beverages, rather than raw water, which could be tainted with harmful microorganisms and other parasites, had a longer life expectancy. It's interesting that multiple groups of people came up with the idea of alcohol independently. Evidence of the oldest booze dates to 7,000 BCE in China. The Caucasus (the area between the Black Sea and the Caspian Sea) began fermenting wine around 6,000 BCE, whereas the Sumerians were brewing beer in southern Mesopotamia (south-central Iraq) around 3,000 BCE. While the agave plants that are used to produce tequila today were also utilised by the Aztecs to create pulque, the Incas brewed chicha, a corn beer. There is evidence that fermented drinks had a vital social function in a variety of contexts, facilitating interaction between groups and perhaps a connection with the afterlife.

However, returning to the Agricultural Revolution, whether it emerged either once a hundred times, it evolved first in the Fertile Crescent, a vast territory stretching from the eastern Mediterranean Sea to Iran and the Levant.

According to recent excavations in certain historical locations, hunter-gatherers began harvesting and growing seeds from wild grains and legumes such as wheat, barley, and lentils as early as 13,000 years ago! Over the course of a few thousand years of cultivation, the wild versions of these plants transformed into new, domesticated species that were simpler to cultivate and sow, making agriculture more productive and efficient.

In 2009, archaeologists Nicholas Conard and Simone Riehl of the University of Tübingen in Germany collaborated with researchers from Iran's Center for Archaeological Research (ICAR) in Tehran, most notably Mohsen Zeidi, an experienced ICAR excavator, to work in the farming village of Chogha Golan in the Zagros Mountains of western Iran. Archaeologists from Iran found the settlement roughly 15 years earlier but never fully excavated it. The researchers discovered considerable evidence of plant processing in the settlement while digging in 2009 and 2010, including mortars, pestles, and millstones. The excavation also uncovered a significant number of scorched plant remnants (almost 21,000 separate pieces).[1]

Radiocarbon dating of the archaeological sites (about eight metres deep) revealed that Chogha Golan was continuously inhabited between about 12,000 and 9,700 years ago, allowing Riehl and colleagues to track the use of plants during that time period. They discovered, for instance, that the residents of Chogha Golan began farming wild barley, wheat, and lentils more than 11,500 years ago, and that cultivated wheat arose around 9,800 years ago, almost as early as in locations to the west. According to the experts, the growth of agriculture in Chogha Golan and the fertile eastern

Crescent was an independent event that paralleled developments farther west.[1]

These findings, according to Roger Matthews, an archaeologist at the University of Reading in the United Kingdom, support the theory of numerous agricultural beginnings. The findings are consistent with recent DNA research that demonstrate that domesticated plants and animals have multiple origins. Furthermore, it confirms work done by Matthews and other British and Iranian archaeologists at another Iranian site called Sheikh-e Abad, where it appears that wild goats were herded, penned, and tamed (probably began around 9000 BCE). In general, the transitional stage between wild and domesticated plants corresponds to the transitional periods in plant utilisation observed in Chogha Golan.[1]

We now clearly understand that agriculture evolved in other remote areas of the world not as a result of Middle Eastern agriculturalists disseminating their revolutionary lifestyle, but all completely independently. In other words, in about 4500 BCE, in the south of Mexico, Central America, mankind domesticated maise and beans without knowing anything about pea and wheat farming in the Middle East. Similarly, in roughly 3500 BCE, South Americans learned how to plant and grow potatoes and tamed wild animals like llamas, completely unaware of what was going on in Mexico or the Levant. Likewise, around 7400 BCE, humans began domesticating rice, millet, and pigs in China's Yangzi River Valley region without having any clue of the indigenous inhabitants of New Guinea's domestication of sugar cane and bananas.

Agriculture spread far and wide from these original foci. By the first century CE, the great majority of people on the planet were farmers. But now, given the fact that the Agricultural Revolution had several different and independent historical roots everywhere in the world, an important and intriguing question has arisen: why have agricultural revolutions occurred in the Middle East, Central America, and China but not in Australia, Alaska, or South Africa?[0]

The simple answer is that most species of plants and animals cannot be completely domesticated. Approximately 11,000 years ago, humans realised that there was a better location for certain creatures than the point of a spear. *Homo sapiens* began enticing them into our settlements, gradually shaping and changing their natures to fit our food, labour, and companionship needs. *Homo sapiens* experimented with domesticating various animals over the ages. Yet, only a select few (most notably the cow, goat, sheep, chicken, horse, pig, and dog) have proven to be so utterly important that they have proliferated practically everywhere humans do. But why only certain creatures? Why not the rhino, tiger, zebra, or any of the thousands of other seemingly suitable creatures that didn't make the cut?[3]

According to the evolutionary physiologist and geographer Jared Diamond, in his acclaimed book *Guns, Germs and Steel*,[2] there are six criteria that animals must meet for domestication. In general, many species come close, but very few fit the bill.[3]

To begin with, domestic animals cannot be fussy eaters and must be capable of finding enough food in and near human settlements in order to survive. Herbivores like cows and sheep must be capable of roaming on grass and consuming our excess grain supply. Carnivores like dogs and cats must be willing to eat human waste and leftovers, as well as the vermin that such morsels attract. Second, only creatures that mature swiftly in comparison to the human life span are worthy of consideration. Humans couldn't afford to waste too much time and energy feeding and caring for an animal until it was mature enough to be put to work or slaughtered.[3]

Third, domesticated animals must be willing to reproduce in captivity. Territorial species, such as antelope, gazelle or impala, cannot be kept in crowded enclosures during reproduction. Even though the ancient Egyptians valued pet cheetahs, the giant cats won't reproduce without extensive wooing rituals, including long-distance sprinting together (hence, they were never domesticated). Fourth, domesticated animals must be docile by nature. Cows and sheep are ideal examples because they are typically laid-back. The African buffalo (*Syncerus caffer*) and the American bison (*Bison bison*), on the other hand, are both unpredictable and dangerous to humans. Similarly, while closely related to horses, zebras are significantly more violent, which may explain why zebras have only been domesticated in rare instances. Some evolutionary biologists believe that docility is not a criterion for domestication because many pets are descended from violent animals, such as the dog (wolf).

Fifth, when frightened, they may not panic and escape. This eliminates most deer and gazelle species, which have a volatile temperament and a strong leap that allows them to flee easily. Sheep, while panicked, have a herd instinct that keeps them close together when they are scared. Finally, except for cats, all big and successful domestic animals follow a social structure characterised by strong leadership, which has made it simple for us to train them to recognise their human caregiver as the pack leader.

A lot of researchers have said that the Agricultural Revolution was a great leap for *Homo sapiens* and made a case for human ingenuity as the source of all growth and development. Humans eventually developed the intelligence to decipher nature's mysteries, allowing them to domesticate a wide variety of animals and successfully grow crops. As soon as this occurred, they gladly abandoned the dangerous and arduous lifestyle of hunter-gatherers in favour of the safe, comfortable existence of farmers.

Humankind's Greatest Fraud

However, American historian Jared Diamond claims that this tale is a fantasy and far from reality.[0] He states that there is no evidence that humans became more intelligent with time, let alone intelligent enough to crack nature's secrets. According to him, since their lives depended on a broad and profound knowledge and experience with the plants they gathered and the animals they hunted, foragers understood the mysteries of nature long before the Agricultural Revolution. Rather than ushering in a new period of ease, the Agricultural Revolution left agriculturalists with lifestyles that were typically more difficult, demanding, and unsatisfying than those of foragers. Hunter-gatherers spent their time in more interesting and varied ways, and they were less likely to die of famine or disease.

The Agricultural Revolution did increase the amount of food available to mankind, but this greater amount of food did not convert into a healthier diet, a more tranquil lifestyle, or more peaceful times. Instead, it resulted in population surges and wealthy elites.[7] The ordinary farmer put in more effort than the average forager and received a worse diet and lifestyle in return. As both Diamond and Harari claim, "the greatest deception and mistake in human history was the Agricultural Revolution".[2]

According to researchers from Washington State University and 13 other schools, the arc of prehistory leans toward economic inequality. The researchers observed increasing income inequalities with the emergence of agriculture, notably the domestication of plants and big animals, and increased social organisation in the largest study of its type. According to the results of Tim Kohler, principal author and Regents professor of archaeology and evolutionary anthropology at Washington State University, published in the journal *Nature*, the Agricultural Revolution had profound implications for contemporary society, as inequality repeatedly leads to social disruption, if not collapse.[4]

The research gathered information from 63 archaeological sites or clusters of sites and assigned Gini coefficients (specific measures of inequality devised more than a century ago by the Italian statistician and sociologist Corrado Gini) by comparing property sizes within each region. In principle, a society with complete wealth equality would have a Gini coefficient of 0.0, whereas a society with complete wealth concentration in one family would have a Gini coefficient of 1.0.[4]

4) Hunter-gatherers constructed huts and inhabited them for extended periods of time in areas with rivers, lakes, and abundant game. These mobile communities typically constructed shelters using impermanent building materials, or they may use natural rock shelters, where they are available.

The researchers discovered that hunter-gatherer communities had minor wealth inequalities, with a median Gini coefficient of 0.17. Their mobility would make accumulating wealth, much alone passing it down to future generations, impossible. Horticulturists (low-intensity, small-scale farmers) had a median Gini of 0.27, and the median Gini coefficient in larger agricultural societies was 0.35. To the researchers' astonishment, inequality increased in the Old World while plateauing in the New World. The researchers and Kohler credit this to Old World cultures' capacity to literally exploit huge tamed beasts such as cattle, horses, and water buffalo.

But who was ultimately responsible for this? Neither kings nor farmers. According to Harari, "*Homo sapiens* did not domesticate wheat. It domesticated us".[0] Just before the

5) In China, around 7400 BCE, humans began constructing artificial defences and domesticating rice, millet, and pigs in China's Yangzi River Valley region. Mobile hunter-gatherers built and inhabited shelters that functioned as campsites, whereas stationary farmers resided in homes or residences that served as the foundation of communities.

Agricultural Revolution, plants like wheat were just wild grass, one of many, restricted to a small range in the Middle East. However, within just a few short millennia, it grew worldwide and became one of the most successful plants in history according to the basic evolutionary principles of survival and reproduction.

Consider, for example, the United States, where in 2008, around 260,000 km² of wheat was sown (larger than the United Kingdom), of which approximately 226,000 km² was harvested respectively.[5] Take a moment to comprehend this enormous size and imagine a wheat field the size of the United Kingdom. Land's End to John o' Groats is the traversal of the whole length of the island of Great Britain between two extremities, in the southwest and northeast, where experienced off-road walkers take about two to three months to walk. Worldwide, wheat covers about 2.25 million km², which is a field ten times the size of Great Britain. To walk across this field will take you years. This is why wheat is so fascinating: it all once came from wild grass growing on limited small fields in the Middle East.

Wheat domestication, on the other hand, was not simple. Wheat demanded a lot among agriculturists because of its "pickiness." Wheat struggled to grow near rocks and stones, so farmers broke their backs clearing the fields all day. Wheat, in addition to rocks and gravel, did not like sharing space, nutrients, and water with other plants; therefore, agriculturalists laboured hard days weeding under the boiling hot Sun. Wheat also became ill, so farmers had to keep an eye out for pests. Herbivorous rodents, in addition to these little creatures, regularly attacked wheat, so farmers constructed fences and kept an eye over the fields. Wheat was hungry and thirsty, so humans dug irrigation canals, dragged heavy buckets from the well to water it, and gathered animal faeces to feed the soil where it grew. Furthermore, because this new agricultural lifestyle required so much time and effort, humans were obliged to stay permanently next to the fields, fundamentally altering humanity's way of existence.

Keep in mind that the bodies of *Homo sapiens* were not evolved for such duties. Considering the evolution and origin of our species, it was designed for running, jumping, and climbing, not plodding through a field all day. Human spines, knees, and necks eventually paid the price. Furthermore, this drastic diet shift had a direct impact on human teeth. Numerous studies have been conducted to determine how the drastic dietary and food processing changes linked with the beginnings of agriculture affected the overall health of Neolithic *Homo sapiens*.[7]

By analysing teeth from Neolithic samples, paleoanthropologists have discovered an overall pattern of worsening dental health among Neolithic people compared to their hunter-gatherer forefathers. Agriculture has been linked to tooth shrinkage, crowding, a rise in caries, and an increase in the prevalence of the periodontal disease. Archaeological evidence reveals that throughout the Neolithic, humans adopted new methods of food preparation, such as grinding stones and cooking in ceramic containers.[10]

High levels of physical activity and stress enhance bone density, which is corroborated by a pattern of decreasing skull size and shape in Neolithic samples, as well as dental microwear data. Due to their more mobile and active lifestyle, Palaeolithic hunter-gatherers most likely had larger skulls than Neolithic people.[9] It is well acknowledged that masticatory forces govern craniofacial growth, and the stress is mostly caused by differences in the mastication action caused by food consistency.

Foods became softer and easier to chew and digest once new preparation and processing processes were developed. This shift in masticatory function led to the general "gracilisation" of the human skull over time, resulting in a smaller human face with smaller jaws and teeth. Face reduction had a negative impact on human dental health since human teeth did not shrink proportionally to the jaw, resulting in crowding. Dental crowding is an issue because it causes small areas between teeth where germs may thrive. These oral bacteria can contribute to plaque buildup and develop caries, an infectious condition of the mouth.[6]

Crowded conditions, along with human settlements near (domesticated) animals, led to high infectious disease rates (this was many because animals were kept close and inside dwellings in many early agricultural villages). Because of this close proximity, several infectious diseases were able to spread from animals to humans. In both animals and humans, contaminated water sources and intimate contact with human waste encouraged parasite infection.

Many early agricultural centres relied on one to three crops and consumed far less meat than their hunter-gatherer forefathers. Grains like barley, wheat, and millet, as well as rice and maize, were significant sources of income for early farming societies.[11] Lower dietary variety also means a reduced diversity of nutrients in diets. Furthermore, reliance on plant foods over meat limits the consumption of zinc, vitamin A, and vitamin B12, which are exclusively found in animal diets.[12]

In addition to less wealth equality, the deteriorated diet and physique, it also didn't provide more economic security. Do you remember the previous chapter, where the omnivorous *Homo sapiens* thrived on a wide variety of plants and meat? Because these hunters and gatherers relied on dozens of species to survive, they were able to adjust when one of the species became scarce by collecting and hunting more of the other species. Yet, farmers did not have access to such a diverse range of food sources. The agricultural economy relied on environmental variables and was built on a cyclical output cycle that included extended months of cultivation followed by brief peak times of harvests. A lack of rain, pests, and other causes might all lead to a failed harvest, which frequently meant starvation.

Farmers also had more things and needed fertile land to farm on. Loss of pasture to raiding neighbours might be the difference between life and death; thus, there was little opportunity for wiggle room. The early farmers were just as aggressive as their foraging counterparts (if not more so).[0] Many anthropological and archaeological research show that human violence was responsible for roughly 15% of deaths in primitive farming communities with no governmental structures or weak "fictional glue". Violence accounts for 30% of deaths in one agricultural tribe group, the Dani, and 35% in another, the Enga, in modern New Guinea. In Ecuador, maybe half of the adult Waoranis die violently at the hands of another human![13]

So, how did wheat and other "picky" crops convince Sapiens to sacrifice a relatively comfortable life in return for a more "bitter" existence? What did it give in exchange? It provided nothing to us as individuals. However, it did benefit *Homo sapiens* as a species: growing wheat supplied significantly more food per unit of territory, allowing *Homo sapiens* to proliferate at an exponential rate. However, the essence of the Agricultural Revolution, according to Harari, was the ability to keep more people alive under adverse conditions; "the Agricultural Revolution was a trap."[0]

Giving up the nomadic lifestyle allowed women to bear children on an annual basis. Extra help was desperately needed in the fields. However, the new mouths rapidly depleted the food surpluses, necessitating the planting of additional farms. When human began to dwell in filthy, disease-ridden villages, children received more grain and less breast milk, and each youngster fought for porridge with more and more siblings, infant mortality skyrocketed. In most agrarian communities, one out of every three children perished before the age of 20. Despite this, the increase in births has been greater than the increase in mortality as individuals have had more and more children.[14]

Over time, things became increasingly difficult. The average person in Jericho from 8500 BCE lived a more difficult life than the average person in Jericho from 9500 BCE or 13,000 BCE. Yet, nobody realised what was happening or where humanity was heading to. Every generation lived like the previous generation, with only exceedingly minor improvements per generation. Ironically, a sequence of "improvements," each meant to make agriculture life easier, added up to a millstone around the neck of these farmers.

But why did humanity make such a catastrophic error in judgement? Perhaps just as humans throughout time have been unable to foresee all the ways in which their actions would play out. Unfortunately, the descendants of these pioneer farmers were unable to change the situation: Due to the gradual nature of the societal transformation, it was several generations before anybody could recall their ancestors' "free" hunter-gatherer past.[15]

The Agricultural Revolution is one of history's most controversial events. Most pleasures become necessities over time, creating additional responsibilities. Humans fall into what is known as "the Luxury Trap" when they grow accustomed to any "luxury" and start taking it for granted.

While some adherents claim the Agricultural Revolution has set humanity on the path to luxury, prosperity and progress, others argue that it has led to destruction. According to them, this was the turning point at which Sapiens rejected their intimate symbiosis with nature and sprinted towards greed and separation. As we will see in the next chapter, whichever way the road led, there was no turning back.[0]

9. THE UNIFICATION OF HUMANKIND

The Agricultural Revolution provided the necessary material and fundamental basis for increasing and strengthening the intersubjective networks. Agriculture made it possible to feed thousands of *Homo sapiens* in crowded cities and thousands of soldiers in disciplined armies. However, the intersubjective webs then encountered a new obstacle: to preserve the collective myths and to organise mass cooperation, the early farmers relied on human brains to process data, which is strictly limited.[0]

Nonetheless, agriculture enabled populations to grow so dramatically and quickly that no complex agricultural society could ever support itself if it switched back to hunting and gathering. Before the Agricultural Revolution began about 10,000 BCE, the world population was estimated to be about a million. Despite the fact that the population of hunter-gatherers had roughly doubled by the beginning of the first century CE (mostly in Australia, America, and Africa), the world's 250 million farmers outnumbered them.[1]

Human societies grew larger and more sophisticated after the Agricultural Revolution, as did the imagined structures that sustained social order. Myths and fiction condition humans to think in certain ways, behave in specific ways, desire certain things, and follow certain norms from birth. As a result, they constructed artificial impulses that allowed millions of strangers to collaborate productively. This network of artificial impulses is referred to as culture.[0] The "creation" of artificial instincts was crucial. Yet, the underlying problem was that *Homo sapiens* evolved in small groups for millions of years of a few dozen individuals. Therefore, the few millennia that separated the Agricultural Revolution from the advent of villages, cities, kingdoms, to eventually empires was not enough time to allow our instinct to develop mass cooperation.

The Rise of Farmers

Most farmers were settled in one place permanently, with only a few herders travelling from place to place. Most people's territories have diminished and become increasingly smaller as a result of their decision to settle down. Hunter-gatherers of the past often occupied large, sprawling territories spanning tens to hundreds of square kilometres. However, a farmer's world consists of his or her home, field, and surrounding hills and streams. As a result, the ordinary farmer felt a deep sense of belonging in his home, which had far-reaching effects, both mentally and physically. A far more egocentric person was reflected in the individual's commitment to the home and isolation from neighbours.

When compared to the vast natural expanses used by ancient nomads, the new agricultural lands were not only significantly smaller but also far more artificial. Hunter-gatherers had made little planned modifications to the landscape they wandered. Farmers, on the other hand, established human settlements by carving off small islands from the surrounding woods. They logged wood, dug canals, levelled land, constructed homes, ploughed rows, and set forth orchards. The resulting manmade environment served primarily human needs, including those of their domesticated plants and animals, and was frequently surrounded by walls and hedges. Farmers took every precaution to thwart the intrusion of wild animals and peculiar weeds. If any such invaders managed to get inside, they were promptly kicked out or wiped out if they persevered. Tough defences were constructed all the way around the home.

Eventually, humans had a hard time leaving their manufactured territory since doing so posed a significant threat to their possessions, such as houses, farms, and food stalls. In addition, they began to accumulate and hoard more and more stuff as time went on. Although today's eyes may roll at the simple lifestyle of ancient farmers, it's important to remember that an average ancient farmer family held more artefacts than an entire foraging tribe.

In one of his lectures, Harari claims that because we are the only species capable of both flexible and massive cooperative behaviour, *Homo sapiens* have absolute power and control over the world. If you recall, there was a plausible explanation in earlier chapters for why Neanderthals weren't able to successfully defend themselves against Sapiens: They were physically superior to the creative, cooperative *Homo sapiens*, but they were eventually swamped because of our species' outstanding reproductive success and capacity to increase our threshold of organised collaboration.

Now, of course, there are other species that can collaborate in enormous numbers, such as the social insects, the bees, and the ants. Ants fight each other in practically every part of the world. Natural selection forged their weapons. Some of them are well armoured, while others feature lethal stingers or razor-sharp mandibles. For decades, researchers have known that ants utilise pheromones, which are tiny chemicals, to communicate with one another. The most well-known example probably involves the pheromone trails left behind by insects as they walk.

The Argentine ant (*Linepithema humile*) wields power over the largest empire ever formed by an ant species, spanning many continents and causing millions of deaths in its conflicts. This supercolony of Argentine ants has been so successful that it has spread to many continents and established sister colonies in California, Europe, Japan, New Zealand, and Australia. By this measure, their civilisation is the greatest in the world.[6] Nonetheless, what sets *Homo sapiens* apart is the fact that these social insects lack the flexibility to collaborate in similar ways to how humans can. Their cooperation is very rigid: almost everything an ant colony needs to survive is written in their DNA, so there's really just one way they can operate.

Empires generate huge amounts of information. They have to keep account of transactions, taxes, laws, inventories, military, traditional events, and many other things. Yet, for millions of years, *Homo sapiens* could only store this information in their brains. However, firstly, the brain capacity is limited; Secondly, when humans die, their brains, and thus the stored information, dies with them; Lastly, and most importantly, our brains have evolved to store and process only particular types of information. Our brains have evolved in the context of gruesome survival, making recognising and remembering shapes, faces, relations of several dozen band members, colours, smells, and behaviour patterns of numerous plant and animal species essential. So consequently, evolutionary pressures have adapted our brains to store immense quantities of social, botanical, zoological and topographical information.

But, with the rise of the Agricultural Revolution, particularly complex societies began to appear, making an entirely new type of information vital: numbers. Yet, because of evolutionary pressures, human brains have never evolved to store and process numbers (which is arguably the reason why we can't truly understand and appreciate big numbers). Yet in order to maintain a complex society, kingdom or empire, mathematical data was crucial.

However, the scale and complexity of human societies were greatly hampered by this mental restriction. After a certain point in time, it became essential to store and process massive volumes of mathematical data because of the increasing population and wealth of a given community. However, it all came crashing down when *Homo sapiens* just couldn't comprehend the information anymore. Because of this, human social networks and societies were small and simple for thousands of years during the initial stages of the Agricultural Revolution.[0]

The Power of Writing

But the first to overcome this core problem were the ancient Sumerians, the earliest known civilisation in the historical region of southern Mesopotamia (modern Iraq). The Sumers emerged during the Chalcolithic and early Bronze Ages between 6000 BCE–5000 BCE and are one of the cradles of civilisation in the world (along with ancient Egypt, Elam, the

Caral-Supe civilisation, Mesoamerica, the Indus Valley civilisation, and ancient China). Farming near the Tigris and Euphrates rivers provided the Sumerians with a surplus of grain and other crops that allowed them to establish urban settlements.

Yet, between 3500–2000 BCE, a Sumerian genius invented a way to store and process information outside the brain, a method that helped overcome the mathematical constraints of the human brain and opened the way for the development of cities, kingdoms, and empires: writing.

The Sumerian writing system combined two types of signs, which were pressed in clay tablets. They used decimal and sexagesimal systems (for example, 600 is ten sixties), so the Sumerian lexical numeral system is sexagesimal with six and ten as a sub-base. Numerals and composite numbers are as follows: 1–11, 20, 30, 40, 50, 60, 600, 1,000, 3,600.[7] Their six-base system resulted in various useful innovations, such as the 24-hour day and the 360-degree split of a circle. Other signs indicated animals, goods, territories, dates, individuals, and so on. When used together, these two types of signals allowed the Sumerians to store and maintain an enormous amount of information, considerably more than could be stored in a single mind or even in a long DNA sequence.

However, in those times, writing was laborious and only used for recording critical information, such as statistics. If you go back to our ancestors' oldest writings, you won't find any great wisdom: According to the most likely "contemporary" interpretation of the phrase "29,086 measures barley 37 months Kushim", this suggests that a total of 29,086 measures of barley were received over the period of 37 months.

6) Sometime around 2600 BCE, in the ancient Sumerian city of Shuruppak, this Sumerian contract for the sale of a field and a house was made.

Yet with this "simple" partial rather than a whole script, the Sumerian kingdom and its metropolis prospered. Around 3100 BCE, at its peak, Uruk, one of Sumer's major urban cities, played host to as many as 80,000 people. The population of Sumer was probably anywhere between 800,000 and 1.5 million, taking into account the neighbouring cities and the enormous agricultural population (for reference, the world population at that time was around 27 million).[8]

The pre-Columbian Andean culture, for example, utilised a different script than Sumerian. In contrast, the Andean script was never properly developed because its users were content with the limits of their writing system and saw no need for improvement. Interestingly, it was not written on physical objects like clay tablets or paper. Instead of using ink, it was written by attaching knowledge onto quipus, which are colourful cords. Large quantities of numeric

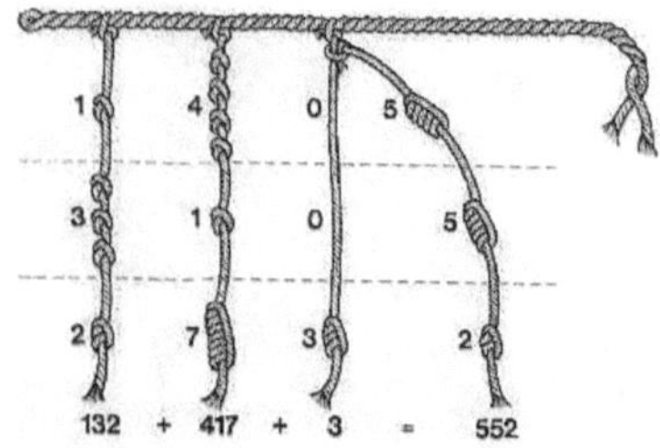

7) Although quipus was not a form of writing, it served as the mnemonic device by which the Incas sent messages, recorded their laws, and decided the fate of conquered territories.

information on property ownership, tax collection, and other topics might be recorded by tying different knots on different coloured ropes.[9]

Under the Inca Empire, also known as Tawantinsuyu by its people, quipus flourished and were an integral part of the economic functioning of towns, kingdoms, and empires for hundreds, maybe thousands of years. Beginning in the early 13th century CE, the Inca Empire dominated the Andean region, marking the end of an era spanning thousands of years of culture in the region. Scholars have identified at least five "pristine" indigenous non-derivative civilisations around the globe, the Andean civilisation being one of them. The empire included much of what is now Chile, Argentina, and Colombia, in addition to modern-day Peru, Ecuador, and Bolivia. Tawantinsuyu's highest population is estimated to have been anything from 4–37 million people (however, most estimates place the population between 10–12 million).[10]

In spite of the fact that the Inca used quipus to keep meticulous census records, virtually all of them eventually fell into disuse, crumbled, or were destroyed by the Spanish. Yet, quipus were so reliable that the Spanish utilised them to manage their new kingdom in the early years after the conquest of South America.

Cuneiform, the entire script based on the Sumerian system, was developed between 3000 and 2500 BCE. Cuneiform, originally used just for writing Sumerian, eventually expanded far and wide as it was altered to document a number of other languages. The majority of the cuneiform record dates to after 2400 BCE and is composed of writings written in Akkadian. A second complete Egyptian script, hieroglyphics, was created at about the same period. China developed a full script around 1200 BCE, whereas Central America did so around 1000 BCE. Eblaite, Elamite, Hurrian, Luwian, and Urartian are a few more languages that have sizable cuneiform corpora. Cuneiform-like symbols may be found in the Old Persian and Ugaritic alphabets (however, they are unrelated to the cuneiform logo-syllabary proper).

With the advent of the complete writing system, *Homo sapiens* began to compose a wide variety of literary works, including poetry, romances, plays, and histories. Despite these works, storing vast quantities of mathematical data remained the primary function of writing. Yet, as more was written and administrative archives expanded, additional difficulties surfaced: the retrieval of information.

Remembering things is a speciality of the human brain. All of our memories are freely linked to one another in our brains. What you ate today, how many cups of coffee you drank, and the last movie or sports game you watched may all be easily retrieved from memory without any time spent looking them up. However, using quipu cords or clay tablets to retrieve the same specific information is vastly different and far less efficient. The use of writing has subtly changed our way of thinking and perceiving the world.

Ironically, it turned out to be far more challenging to develop efficient systems for storing, organising, and retrieving written materials than it had been to develop the written word itself.[0] What truly characterised ancient empires such as Sumer, Akkadian, Inca, Egypt, China, Greek, and the Roman Empire was their excellent recordkeeping, cataloguing, and retrieval techniques.

The latter empire cannot be underestimated. The origins of Roman numerals date back to the 8th to 9th centuries BCE, roughly contemporaneous with the foundation of ancient Rome. The Roman numeral system was derived from ancient Etruscan numerals (which were itself derived from Greek Attic symbols). In comparison, the Romans wrote I, V, X, L, C, D, and M for 1, 5, 10, 50, 100, 500 and 1,000. Similar to Etruscan, Roman numerals are written using addition and subtraction. Addition notation indicates that the individual values of a Roman numeral are added to create the total. Therefore, Romans would write the number 37 as 10 + 10 + 10 + 5 + 1 + 1 (XXXVII).

However, the Roman system had a few flaws, including the absence of a zero sign and a method for numbering over several thousand that did not include placing lines around numbers to denote multiples. Nonetheless, this did not prevent the ancient Roman intellectuals and architects from constructing a mighty empire. The administration of a complex society and economy, as well as the construction of enormous structures such as the road system, Colosseum, Pantheon, and aqueducts, needed a high level of mathematical proficiency.[11]

There are several hypotheses on the design of the ancient Etruscan and Roman numbers. One such idea said that they originated from the tally sticks herders used to count their livestock. Shepherds used to notch their sticks; hence, I became a single unit, with a double-cut every fifth (or V) and a cross-cut every tenth (X). The other central theory was that these were allusions to hand signals, with I, II, III, and IIII equating to single fingers and V being exhibited with the thumb extended and fingers interlocked. The numerals 6 to 9 were represented with a V on the one hand and I, II, III, or IIII on the other, while the number 10 (X) was represented by crossing the thumbs.[11]

Interestingly, despite its flaws and limitations, the Roman number system lasted longer than the empire itself, continuing in widespread use until the 14th century CE, when it was replaced by the Arabic system (which had been introduced to Europe in the 11th century CE).

The Arabic numeral system, established during the 9th century CE, is an extraordinary script. You'll most often see these ten digits (in the order 0, 1, 2, 3, 4, 5, 6, 7, 8, and 9) while dealing with decimal numbers (the primary reason the source reference numbers start from 0 rather than 1). They are also used for writing numbers in other systems, such as computer symbols, trademarks, and registration plates, as well as for writing numerals in other systems, such as octal. Particularly when contrasting with Roman numerals, the phrase is commonly taken to mean a decimal number. Despite their widespread association with the Arab conquest of India, many argue that the Hindus were really the ones who initially came up with the idea for what we now call Arabic numbers. However, the proper foundation of contemporary mathematical notation was laid when several more signs were added to the Arab numerals (such as multiplications, subtractions, additions, and so on).

Interestingly, the history of mathematics dates back thousands of years to ancient Egyptian, Mesopotamian, and Indus Valley civilisations. Ancient civilisations utilised mathematics for applications like measuring land, designing and constructing buildings, and documenting astronomical observations. They created systems of numeration and arithmetic that allowed them to conduct simple calculations and solve straightforward issues.

In ancient Greece, mathematics took on a more theoretical and abstract shape. The mathematicians of this period, such as Pythagoras, Euclid, and Archimedes, created new concepts and procedures that established the groundwork for contemporary mathematics. The Greek mathematician Pythagoras discovered that the square of the hypotenuse is equal to the sum of the squares of the other two sides of a triangle with one of the angles at 90 degrees: $a^2+b^2=c^2$. Pythagoras' theory, widely renowned as the Pythagorean theorem, may be traced back to ancient Babylon and Egypt (starting around 1900 BCE), where comparable concepts were utilised to solve issues with right triangles. The partnership was shown on an ancient Babylonian tablet known as Plimpton 322, which dates back 4,000 years. Until Pythagoras stated it explicitly, the relationship was not well known.

From the 14th century to the 17th century CE, algebra, geometry, and calculus continued to develop throughout Europe during the Renaissance, along with other branches of mathematics. Mathematicians such as Leonardo da Vinci,

Galileo Galilei, and Isaac Newton developed discoveries that transformed the field.

New concepts and approaches have emerged in disciplines such as topology, number theory, and mathematical physics as mathematics have continued to expand and develop in the modern era. Mathematics is an essential component of modern science, technology, and industry, having applications in domains such as cryptography, computer science, and finance. It allows for the rapid and efficient storage, distribution, and processing of any information that can be converted into a mathematical script.[0] As a matter of fact, due to the worldwide nature of mathematics, it may serve as a universal language and is arguably the most widely spoken language in the world (Yet, linguists who do not regard mathematics to be a language claim that it is written rather than spoken).

However, coming back, another essential question boils down to why the ability to write, store, and effectively retrieve mathematical data was crucial to the maintenance of a complex empire: Since an imagined order is always in danger of collapse because it depends upon myths, and myths vanish once people stop believing in them, how did humans organise themselves in mass-cooperation networks, despite their lack of the biological instincts necessary to sustain such enormous networks?

The Economic Order

This profound question was resolved in the first millennium BCE when three orders emerged with the potential to unite all of humanity under a single set of laws.

The first universal law to bring about a unification of humankind was based on money.[2] Complex commercial systems can only function with some sort of monetary exchange. To be a successful blacksmith in an economic society, one need only be knowledgeable of the market value of various tools and weapons. You shouldn't try to remember how much one piece of equipment or weapon costs in comparison to other necessities like food or clothes. Specialists in apples no longer have to seek out blacksmiths with an appetite for apples since everyone nowadays is interested in making a profit. In many ways, this is the most essential aspect of it. Simply said, you can buy or sell anything you want with money since everyone wants money. The blacksmith will cheerfully take your money since he can always buy the things he really wants with it, such as food and clothing. That's why money is such a convenient means of exchange; it can be used to buy or sell practically anything, wherever.

In addition to serving as a medium of exchange, ideal forms of money may also be used as a convenient means of wealth storage for humans. Time and beauty are two examples of commodities that cannot be stored. For instance, fruit doesn't preserve very well. Some items survive longer, but they're bulky and expensive to store and preserve. Grain, for instance, has a long shelf life, but storing it needs massive warehouses that are secure against rodents, mould, water, fire, and theft. These issues can be remedied by exchanging paper currency, digital currency, or exotic shells for one another. However, coins are non-perishable, repel vermin, are fireproof, and are small enough to store in a safe.

Because they can be quickly and cheaply converted, stored, and transported, coins have played a crucial role in the development of complex economic networks and dynamic marketplaces. Without monetary incentives, commercial networks and markets could not have grown to their current scale or been as dynamic. When humans put their faith in a made-up story, they are more likely to help complete a task for an unknown party.

All forms of currency start with trust. In this sense, money is a mutual trust system, and not just any mutual trust system: it is the most pervasive and effective mutual trust system ever created. The foundation of this relationship is a network of political, social, and economic relationships that stretch back many millennia. Why should you trust an unusual shell, a gold coin, or a dollar bill? For the simple reason that your family and friends trust them. Because you believe them, other *Homo sapiens* do too. And we all believe in them because our government does too, as evidenced by the "lovely" tax payments it sends our way.

Since archaic humans lacked this trust, it was crucial to define "money" as commodities with true intrinsic worth. One such example is the Sumerian barley money, the world's earliest currency. It originated in Sumer (about 3000 BCE) with writing and under similar circumstances. Similar to how writing developed to meet the needs of expanding administrative tasks, barley money arose to accommodate expanding commercial exchanges.

Although barley had inherent worth, it was challenging to convince humans to use it as money rather than a commodity. Now imagine you went to the grocery store with a bag full of barley and tried to buy some bananas and a box of chocolates. The shopkeepers would probably think you're crazy. It was far simpler to establish confidence in barley as the initial kind of money because of its intrinsic biological worth: for human consumption. However, barley presented logistical challenges both during storage and shipping (Imagine having to store, move, secure and buy an expensive dinner, luxurious vacation, pricey automobile or a massive mansion with carts full of barley). Therefore, the true innovation in the history of money was the widespread acceptance of currencies that had no intrinsic value but were simpler to keep and transport.

According to Greek historian Herodotus, western Anatolia is the site where history's first coins were forged (around 640 BCE) by the order of King Alyattes of Lydia (modern-day Turkey). These coins were struck with a unique hallmark and were minted from normal gold or silver. Two things might be deduced from the symbol. One, it indicated the weight of the coin's precious metal content; Second, it revealed the name of the government agency that mints and backs the currency. Virtually all modern and ancient coins trace their ancestry back to Lydian ones, such as the shekel.

Shekel is a silver-based ancient Mesopotamian currency. A shekel was first a unit of weight—approximately 11 grammes (0.39 oz)—before becoming an actual currency in ancient Tyre, ancient Carthage, and subsequently ancient Israel under the reign of the Maccabees. The first reported use of the word shekel was in 2150 BCE under the Akkadian Empire, and again in 1700 BCE in the Code of Hammurabi. Like with many other ancient units, however, the shekel had varying values depending on the era, government, and region. Recorded earnings for labourers in the ancient world varied

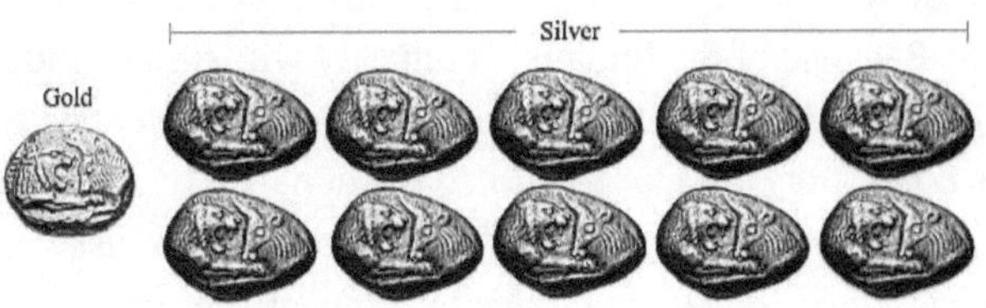

8) Croeseid (Lydian king from 560–546 BCE) bimetallic equivalence: one gold Croeseid of 8.1 grammes was equivalent in value to ten silver Croeseids of 10.8 grammes.

considerably. The Code of Hammurabi (about 1800 BCE) assigns an annual wage of around ten shekels to unskilled labour. Certain documents from the Persian Empire (539–333 BCE) indicate that the monthly salary for unskilled labour ranged from a minimum of two shekels to as much as seven to ten shekels. During the Persian era, a household in a city would have needed at least 22 shekels per year to survive.

However, returning back the significance of coins, compared to unmarked metal ingots, coins offer two major benefits. First, the former required initial transaction-specific weighting. Second, it's not enough to just weigh the ingot. The stamp verifies their true value, so the blacksmith doesn't have to keep a weight next to his register. What's more, the coin's imprint serves as the official seal of some governmental entity, guaranteeing the coin's legitimacy and worth.

Merchants outside of Rome's borders were surprisingly open to accepting denarii as payment due to widespread confidence in the legitimacy of the Roman currency. Roman currency was extensively accepted in Indian markets by the first century CE, despite the fact that the nearest Roman legion was located hundreds of kilometres away. Because of the Indians' familiarity with and confidence in the denarius that when local governments produced their own coins, they closely resembled the denarius, right down to the emperor's picture. Denarius started to be used as a generic word for currency. Arabisation of the word "dinar" occurred at the time when it was being used by Muslim caliphs. Amazingly, the dinar is still in use as the official currency in the countries of Jordan, Iraq, Serbia, Macedonia, Tunisia, and Yemen.

However, the introduction of paper currency took considerable time. In China, during the Tang and Song dynasties beginning in the 7th century CE, the earliest known banknotes were produced. Frankly, it was not globally utilised as money; therefore, it was not an actual currency. However, upon establishing the unified Mongol Empire, Genghis Khan introduced gold and silver coins known as Sukhes and, most importantly, in 1227 CE, the world's first banknote. Around 1260 CE, under the reign of Kublai Khan, the "Deposit Bank" was created in Khar-Khorum, the capital of the Mongol Empire, and banknotes with values of 10, 20, 100, and 1,000 were produced. Eventually, Europeans got introduced to banknotes by merchants and travellers such as Marco Polo and Willem van Rubroeck (Yet, only in 1661 CE did Sweden produce the first banknotes in Europe!). Marco Polo's belief in paper money under the Yuan dynasty (Mongol Empire) was so great that he wrote an entire chapter about it in his book *The Travels of Marco Polo* by describing it as "one of the world's greatest marvels."

Yet, with the birth of this highly effective and functional currency, a global monetary zone was established.

The establishment of a universal monetary zone laid the framework for the future integration of Afro-Asia and, by extension, the entire world into a single economic and political realm. While humans throughout the world persisted in speaking incomprehensible languages, governing themselves in a wide variety of ways, and worshipping a wide variety of deities, they did have a common belief in the value of gold and silver. Without this consensus, global trade networks would never have been established.

Philosophers and academics have been demonising wealth as the root of all evil for centuries. But money is the ultimate test of human tolerance. Money is more tolerant than any language, government regulations, cultural norms, religious beliefs, and social practices. Money is the only trust mechanism ever established by *Homo sapiens* that is universally accepted regardless of factors like race, religion, age, or sexual orientation. With the assistance of money, even humans who don't trust each other may cooperate effectively. Thus, money is based on two such principles. First, money, for one reason, has the potential to work as alchemy, changing the value of land into loyalty, justice into health, and violence into knowledge. Second, universal confidence: two humans, even complete strangers, may work together on any project that involves exchanging money.[0]

Though somewhat true, the philosophers and intellectuals who criticised money were not entirely off-base (money has a dark side). This is due to the fact that monetary investment is made not in people, groups, or sacred values but in monetary systems, which are inherently impersonal despite the fact that they create trust between strangers. Not the stranger or the neighbour, but the coin in their hands has our faith. So, if they run out of coins, we'll lose trust in them. As financial transactions transcend national, religious, and cultural boundaries, the world faces homogenisation into one merciless marketplace.

The Religious Order

But returning to humankind's unifying universal laws, the second was religious: the order of universal solid religions.[3] A belief in a higher, "super" human order or "higher" hierarchical power is at the heart of every major world religion. For this reason, we focus on two primary prerequisites: Religions hold to the existence of a higher order than that which results from human preferences and judgements, and on the basis of this higher order, they construct rules and values that are held to be binding.

Although many faiths have used their influence to shape societal structures, others have not. Two more qualities are needed for a religion to unite a large area populated by many different kinds of people. To begin, it must defend an eternally reliable supernatural order. The second step is to keep convincing everyone of this truth. It ought to have a worldwide scope and missionary zeal.

Both Islam and Buddhism, two of the world's most widely practised faiths, are labelled as "global religions" and "missionary religions" by Christians. This leads many to falsely generalise that all faiths are equally inerrant. Most early faiths were confined to certain regions. Their followers honoured the gods and spirits of their homelands and had no interest in forcing a global conversion. As far as we can tell, the earliest appearance of global and missionary religions dates back to the first century BCE. Like the rise of global empires and the introduction of universal money, their growth was a momentous historical event that sparked widespread change and brought people closer together.

Animism (from Latin: "anima" meaning "breath, spirit, life"), in one form or another, has been the dominant religious tradition throughout all human cultures since our ancestors first migrated out of Africa. Current academic philosophy of religion is eerily silent on the topic of animism, despite the fact that animistic beliefs and practices are nearly universal among indigenous peoples on every continent, and despite animism's critical significance in the early emergence and evolution of human religious thought.

Animists, at the very least, hold that trees, lakes, mountains, thunderstorms, and animals are all sentient entities with whom humans may create social ties. Moreover, many animist societies consider elements of the natural world to be non-human relatives or ancestors from which the present-day members of the civilisation are descended.

For the most part, hunter-gatherers didn't venture outside of an area any bigger than a thousand square kilometres. In order to ensure their own survival, locals had to understand the supernatural order that ruled their area and adapt their ways of living appropriately. It was useless, however, to try to persuade the inhabitants of some far-off land to adopt the same values (the holy mountains of the Alps have no value for those living near the sacred water of the Nile river. Thus, learning and comprehending this hierarchical value was a waste of time for outsiders). As a result, many animist faiths have a narrow geographical scope, privileging the unique features of their respective environments and the peculiarities of their own unique events.

One legend claims that gods rose to power because they offered answers to fundamental human problems. The gods of fertility, eternity, and healing all rose to prominence as plants and animals lost status or their "ability" to speak. For instance, after Uranus was defeated by Cronus, the ancient Greeks fashioned the lovely goddess Aphrodite out of sea foam. Therefore, she has the distinction of being the eldest of the Olympians (raised from the foam is an approximate translation of her name). It was widely accepted that Aphrodite was the superior love and attractiveness deity. As a result of this niche deity, women and men alike fell hopelessly in love with her. Most depictions of Aphrodite see her as a vengeful woman. Throughout Greece, Aphrodite was worshipped, and many priestesses acted as her representatives.

Much of ancient mythology may be seen as an agreement in which people promised eternal allegiance to the gods in exchange for power over the plants and animals. Lambs, alcoholic beverages, and baked goods were commonly sacrificed to the gods for thousands of years following the Agricultural Revolution in the hope of ensuring plentiful crops and healthy cattle.

Stones, rivers, spirits, and devils all had less of a role in the animist system when the Agricultural Revolution began. Even their popularity began to dwindle as humans shifted their devotion to the new gods. If humans lived their whole lives in a very small area, local spirits could be able to meet most of their demands. However, as empires and commerce networks grew in size, citizens needed a way to communicate with authorities that had jurisdiction over a wide area. In order to meet these needs, polytheistic religions emerged (from the Greek "poly", meaning many, and "theos", meaning god). The fertility goddess, the rain god, and the war god were all seen as supreme deities by adherents of their faiths. These deities are prayed to by humans in the hopes that, in exchange for devotions and sacrifices, they will bless the world with favourable weather, military success, and medical well-being.

After Assyria conquered Babylon during the Sargonid period (8th–7th century BCE), Assyrian scribes began using cuneiform to write the name of Ashur, the first religion of ancient Assyrians and Akkadians, and the head of the Assyrian pantheon in Mesopotamian religion. This realm, adorned with jewels and blooming with flowers, had a mythology so ancient that it seemed to have been the progenitor of all others. The Assyrians counted no less than three hundred spirits of Heaven and six hundred spirits of Earth, all of which (along with the rest of their mythology) appear to have been borrowed from ancient Babylonia, the cradle of mythology that later became the inheritance of so many nations in various forms.

Yet, nearby ancient Mesopotamia, around the 18th century BCE, Minoan singers propagated the Greek myths through an oral-poetic legacy; eventually, the mythology of the heroes of the Trojan War and its aftermath became part of the oral tradition. In addition to explaining natural events, cultural traditions, and the secrets of life and death via tales of myths, heroes, and creatures, the enthralling Greek mythology also reveals their polytheistic religion. Greek

9) Fragment of a Hellenistic relief (1ˢᵗ century BCE–1ˢᵗ century CE) depicting the twelve Olympians carrying their attributes in procession.

religion is considered to have originated from the beliefs of the Mycenaean civilisation (which preceded the Minoan civilisation), thriving in Greece from 1600 BCE to 1100 BCE. The Mycenaeans worshipped several deities, including sky gods, earth goddesses, and fertility and agriculture-related deities. By the time of the classical period in ancient Greece (approximately the fifth to fourth centuries BCE), the pantheon of Greek gods and goddesses had formed. It was believed that the gods resided on Mount Olympus, the tallest mountain in Greece, and that they possessed feelings and personalities similar to those of humans. In their honour, temples were constructed, and priests and priestesses acted as intermediaries between the gods and humankind.

However, roughly a millennium later, in 500 CE, like the Mesopotamian and Greek religions, the original polytheistic and pagan Old Norse religion of the Vikings emerged in what is now Denmark. When Christianity gained a foothold in Scandinavia beginning in the 8ᵗʰ century CE, the number of pagans dropped. Yet, this ancient practice preserved Viking culture. The Vikings had two pantheons that continuously interacted with one another, much as many

Viking tribes did (sometimes allied, but often at war). The majority of the gods that the Vikings habitually worshipped were members of the Aesir tribe, who were essentially the "main gods", including Odin, the supreme deity of Norse mythology and the greatest among the Norse gods and the Allfather of the Aesir. The second clan, the Vanir, included the fertility gods and identified Njord, Freyr, and Freyja among their most noteworthy members, who were frequently regarded as honorary Aesir. Despite their animosity, it was vital for the two families to unite their strengths and goals in order for everyone to prosper.

Additionally, many centuries later, on the other side of the world, the Incas (1200-1533 CE) also worshipped many gods, including the creator god (Wiracocha), the sun god (Inti), the moon goddess (Mamaquilla), the thunder god (Illapa), the Earth Mother (Pacha Mama), and a variety of other animistic supernaturals.

Nevertheless, in today's world, polytheism is noted for being part of Hinduism, Mahayana Buddhism, Confucianism, Taoism, Shintoism, and contemporary tribal religions in Africa and the Americas. However, while polytheism made great strides in its ability to replace animism, it was not successful in doing so. Almost all polytheistic faiths maintained a strong emphasis on the supernatural, including demons, spirits, sacred stones, rivers, valleys, and plants. Even though these spirits couldn't compete with the great gods, they met the demands of many commoners.

Prominent deities emerged primarily as a result of elevated human social status. Humans, in the eyes of animists, were simply another species among countless on Earth. In contrast, polytheists interpreted the cosmos as a reflection of the divine connection to humanity. Thus, polytheism elevated not only the gods' but also humans' status.

Polytheism encourages mutual respect amongst faiths. Because polytheists believe in one ultimate and entirely unbiased force on one side, and many partial and biased powers on the other, monotheists have no trouble recognising the existence and efficacy of other gods. Polytheism is widely seen as progressive since "non-believers" were seldom punished. While in power over huge territories, polytheists made no attempt to convert their followers. To its very foundation, the Roman Empire was a polytheistic society and held to a polytheistic religion despite the existence of monotheistic faiths like Judaism and early Christianity within the empire.

Some adherents of polytheistic religions grew so devoted to a single deity that they abandoned the religion's core tenets. Over time, they came to think that their deity was the only god, and that it held all cosmic authority. At the same time, they maintained the view that their deity had preferences and could be used for financial gain. This led to the emergence of monotheistic faiths, whose followers prayed to the supreme being for protection against evil and victory in battle. New monotheistic faiths sprang from polytheism on occasion but remained marginal.

In 350 BCE, Pharaoh Akhenaten of Egypt stated that the god Aten, a minor god in the Egyptian pantheon, was in fact the supreme authority influencing the Universe, marking the beginning of the first known monotheistic religion.

Christianity represented a major innovation. This faith originated as a small Jewish sect trying to convince their fellow countrymen that Jesus of Nazareth was the promised Messiah. Paul of Tarsus, an early Christian leader, reasoned that everyone, not just Jews, should know about Christianity because if the highest force in the Universe had interests and prejudices, and if He bothered to incarnate Himself in the flesh and die on the cross for the salvation of humanity, then everyone, not just Jews, should know about it. Therefore, it is crucial that the great news of Jesus Christ be shared all across the globe.

The triumph of Christianity acted as a model for another great monotheist religion, Islam, which was established on the Arabian peninsula in the 7th century CE. Like Christianity, Islam began with a small minority in a far-flung part of the world. However, in an even wilder and more rapid historical surprise, Islam sprang from the Arabian deserts and conquered an immense empire stretching from Spain to India. The monotheist idea then became central to the development of civilisations everywhere.

One thing all the faiths we've looked at so far have in common is a focus on the supernatural. Some of these natural law religions maintained a theistic tenet, but their gods were held to the same physical laws that governed humans, animals, and plants. One of the oldest and most widely practised religions, Buddhism is exemplary of the natural law religion (Buddhism is a religion without a concept of a one creator God. It is a kind of trans-polytheism that allows numerous long-lived gods, but considers Nirvana to be the ultimate reality).

There were very few monotheists during the beginning of the first century CE. By the early 4th century CE, the Roman Empire had become a Christian imperial state, and missionaries were actively converting the whole known globe. A majority of people in Eurasia and North Africa at the end of the first millennium CE thought they were chosen by a supreme deity. By the turn of the second millennium CE, around 3.3 billion *Homo sapiens* had embraced one of the three monotheistic faiths!

Today, most people blame religion for fostering bigotry, violence, and division. But as a matter of fact, as of today, religion continues to function as the fundamental uniting force of *Homo sapiens* (You could arguably even say that you ultimately exist right now, right in this place, because of the evolution and unifying force of religion). All social structures and hierarchies are precarious because they are imaginary and fictional, and the more the population, the more precarious they become. Throughout history, religion has been instrumental in lending these frail institutions a veneer of supernatural legitimacy. Religions contend that social standards are not the product of random human choice but rather of an ultimate and sovereign force. By making at least certain basic norms unquestionable, society can remain stable.

The Imperial Order

Nevertheless, humankind's last significant unifying order was political: the imperial order.[4] There are two defining characteristics of an imperial political system. To earn such a title, one must first establish authority over a sizable number of distinct peoples, each with its own language, customs, and geographical boundaries. As a result of their cultural diversity and geographical adaptability, empires not only have a unique character, but also make a lasting impact on the course of history. More and more of the world's population and landmass have been brought together under a single governmental umbrella thanks to the success of many empires.

Second, the borders of an empire may expand and contract as needed, and their appetite can never be satiated. They could be able to absorb and incorporate an increasing number of nations and regions while keeping their core structure and identity. While the existing British state, for instance, has distinct boundaries that cannot be crossed without radically changing the country's fundamental structure and identity, a century ago, almost every region on the Earth may have been a part of the British Empire.

It's important to emphasize that an empire is defined not by its origins, political structure, territorial expansion, or population size, but rather by the diversity of its culture and the flexibility of its borders. The creation of an empire is not a necessary consequence of military victory, and size doesn't matter too much either. An empire doesn't have to be large to be powerful. At its peak, the Roman Empire covered far more area and had a much smaller population than modern-day Italy. Or compared to modern-day Mexico, the Aztec Empire was significantly smaller. Yet, the contemporary countries of Italy and Mexico are not the same as their imperial predecessors since they formerly conquered dozens or even hundreds of other lands while the latter did not

It's intriguing to consider how such a human mishmash might exist inside the confines of a little contemporary state. Because there were a lot more different kinds of people living on Earth before, and each of them had a smaller population and required less land than the average person does now, this was achievable. While the land between the Mediterranean and the Jordan River presently has trouble providing for the requirements of only two peoples, it formerly supported dozens of independent governments, tribes, and kingdoms.

Empires had a crucial role in the significant decline in human diversity. The imperial snowplough gradually wiped away the distinguishing characteristics of innumerable peoples, establishing new and far larger groups out of them. Additionally, *Homo sapiens*, like other social animals, have evolved to be territorial creatures. Sapiens split mankind intuitively into "us" and "them." This new imperial vision was passed down from the Persians to Alexander the Great, who in turn passed it down to Greek monarchs, Roman emperors, Muslim caliphs, Indian dynasties, and, eventually, modern-day government leaders.

Empires were critical in combining numerous little civilisations into fewer large ones. Within the bounds of an empire, ideas, people, products, and technology spread more easily than in a politically divided territory. Empires themselves were frequently the ones that actively fostered and disseminated ideas, institutions, customs, and values. One motive was to make their lives simpler. It is difficult to oversee and govern an empire, with each region having its own set of laws, writing system, language, and currency.

Another significant reason empires deliberately disseminated a common culture was to achieve authority. Since the dawn of empires, they have rationalised their actions, whether helping or killing, as necessary to disseminate a higher, stronger and more superior civilisation that benefits the conquered even more than the conquerors.

Cultural ideals propagated by the empire were rarely the sole invention of the rulers. Because the imperial vision is global and inclusive, imperial elites found it relatively easy to integrate ideas, beliefs, and institutions from wherever they found them, rather than obsessively clinging to a particular hidebound tradition: Rome's imperial culture was almost as much Greek as it was Roman. The imperial Abbasid civilisation (the third caliphate to succeed the Islamic prophet Muhammad) was a fusion of Persian, Greek, and Arab influences.

It's tempting to divide humans, civilisations and empires into good or evil. Yet, for the great majority of human history, kingdoms were established on gruesome slaughter, and their control was sustained via tyranny and warfare (pretty evil, right?). Nonetheless, the majority of today's cultures are built on colonial legacies. What does it say about us if empires are by definition evil?

However, it seems improbable that humanity will be able to meet these obstacles without global collaboration. It remains to be determined how such collaboration may be achieved. Nevertheless, as we shall see in the next chapter, civilisations, empires, and kingdoms have all claimed to construct a universal political system for the benefit of all humanity from their inception. Nonetheless, they all lied and ended in total failure. No kingdom was truly universal, and no empire fully served all humankind. Will any current or future power do any better?

10. IMPERIAL HISTORY

Efficient and strong commerce, thriving empires, and strong religions eventually brought nearly every *Homo sapiens* on every continent together to form the global world we live in today. However, this process of growth and unification was not without hurdles. Nevertheless, as we shall see in this chapter, by looking at the big historical picture, the transition from villages, towns, cities, kingdoms, and empires to eventually a single global hyperconnected society was inevitable due to the dynamics of human history.

Based on the work of Matt Baker, PhD, in Religious Studies;[1] Ian Matthew Morris, PhD, British historian and Willard Professor of Classics at Stanford University;[2] George Modelski, PhD, Emeritus Professor of Political Science at the University of Washington,[3] and many more, we will chronologically meander through humankind's imperial history and its major periods and eras. Therefore, it's important to consider that every page in this chapter (and frankly, the entire book) represents a crossroads. A single travelled road connects the past and present, yet other paths branch off into the future. Some of those paths are bigger, straighter, and better designated, making them more likely to be travelled. However, as we shall see, history, or the people who make history, can take unexpected turns.[0]

Let's now enter our fictional time machine and start our journey during the Neolithic period, also known as the New Stone Age or the Stone Age's last division, when farming, animal domestication, and the transition from a hunter-gatherer lifestyle to civilisations began. The temperature on Earth was warmer during the New Stone Age than it was during the Old Stone Age. As we have already covered, no one knows for sure why the Earth warmed, but we do know that the last glacial period ended around 12,000 years ago. As the planet warmed and its climate altered, the Agricultural Revolution occurred.

The New Stone Age

10,000-2000 BCE | Estimated World Population: 4.4 million

Approximately around 7500 BCE, somewhere in modern-day Turkey and Jordan, the most thriving ancient proto-cities, Beidha, Basta and Çatalhöyük, are estimated to have reached a population of 1,000 (Interestingly, as a quick fact, based on Early Neolithic data, the average life expectancy during this time would be about 45 years).[4] Impressively, Jericho, the oldest continuously inhabited city in the world (founded around 9600 BCE), probably also had a population of 1,000 around the same time. To comprehend the fact that the ground beneath Jericho has vibrated to the tune of continuous *Homo sapiens* habitation for more than 11,600 years is astonishing.

If we accelerate several thousand years towards the present, ultimately appearing in 6000 BCE, Çatalhöyük has eventually grown into a 3,000-inhabitant rich, thriving town. Yet, this thriving ancient town did not last long and was entirely abandoned 300 years later.

The first populated settlements in human history to have 5,000 people were Tell Brak (modern Syria) and Uruk (modern Iraq) around 4000 BCE. Although the legacy of Çatalhöyük faded away soon after its zenith, the city of Uruk thrived tremendously and reached 45,000 residents (with 80,000–90,000 people living in its environs), making it the world's greatest metropolis at the time (3000 BCE). Unfortunately, the city's significance faded after 2000 BCE in the backdrop of Babylonia's conflict with Elam.

Nonetheless, it was occupied during the Seleucid (312–63 BCE) and Parthian (227 BCE to 224 CE) eras before being abandoned shortly before or after the Islamic invasion of 633–638 CE. Interestingly, however, if we go back to 3000 BCE, when Uruk reached 45,000 inhabitants, its maximum inhabitants and peak still hadn't been achieved.

The Bronze Age

2000–1200 BCE | Estimated World Population: 72 million

However, let's take great strides in our journey, eventually arriving in 2000 BCE, at the start of the Bronze Age. The Bronze Age occurred in three main areas: Egypt, Sumer (modern-day Iraq), and the Indus Valley (modern-day Pakistan). Unfortunately, since the writing system used in the Indus Valley has yet to be deciphered, we can only read the records from Egypt and Sumer. However, we do know that these three civilisations did trade with one another and that throughout the early Bronze Age, each grew in size and technology.

Sumer arose from a scorching, dry terrain bereft of water to cultivate crops, trees, or building stones. Despite this, its inhabitants built the world's first metropolis, complete with huge architecture and massive populations, primarily out of mud.

Sumer inhabited the southern section of modern Iraq, known as Mesopotamia (Mesopotamia means "between two rivers," referring to the Tigris and Euphrates rivers). Around 5000 BCE, early Sumerians built superb irrigation systems, reservoirs, and dams to reroute river water and harvested vast tracts of previously bone-dry terrain. Agricultural villages like these were sprouting up all over the world.

The Sumerians, however, were the first to embark on the next step: they began to construct multi-story houses and temples out of clay bricks manufactured from riverbed mud. They invented the wheel, which was used to convert mud into everyday items and tools. However, during the early Bronze Age, they also invented money, the sail, astronomy, and mathematics. Sumer, with its superior knowledge and craftsmanship, gave birth to the world's earliest cities, most likely about 4500 BCE.

Priests and priestesses were regarded as nobles at the top of the city's hierarchical order, followed by merchants, craftsmen, farmers, and eventually slaves. The Sumerian empire consisted of independent city-states that functioned as microstates and were tightly connected by language and religious practice (but remarkably, there was no central governmental rule). Uruk, Ur, and Eridu were the cradle of the earliest cities, which were followed by several more. Each city had a king who was a combination of a priest and a monarch, and they periodically fought to acquire additional territory (each city had a guardian deity who was considered the city's founder).

The ziggurat, a temple constructed as a stepped pyramid (home of the guardian god), was the city's largest and most important structure. Sumerians started to spread their influence in 3200 BCE: The potter's wheel moved to chariots and waggons, they made boats out of reeds and date palm leaves, with linen sails that they used to travel great distances by river and sea, and established important trading networks with the burgeoning kingdoms of Egypt, Anatolia, and Ethiopia, importing gold, silver, lapis lazuli, and cedar wood to augment scant supplies.

Memphis, the ancient Egyptian capital, somewhere near the Nile River, was thriving during this vigorous and prosperous trading age, and eventually reached the historic milestone of 60,000 residents. Ancient Egyptian civilisation followed prehistoric Egypt and merged around 3100 BCE.

Additionally, around this time, somewhere very far away, the Xia Empire, better known as the Xia dynasty, emerged in China around 2205 BCE. The Xia dynasty was the first dynasty in traditional Chinese historiography and was established by the legendary Yu the Great, after Shun, the last of the Five Emperors, gave the throne to him. Based on many different and uncertain calculations, the Xia Dynasty ruled between 2070–1600 BCE.

However, by the third millennium BCE, Sumer was no longer the only power around Mesopotamia (nomadic tribes invaded in droves from the north and east). In addition, the Akkadian Empire (founded approximately 2334 BCE) was Mesopotamia's first ancient empire after the long-lasting Sumerians and reigned over Akkadian and Sumerian speakers. These immigrants looked up to the Sumerians,

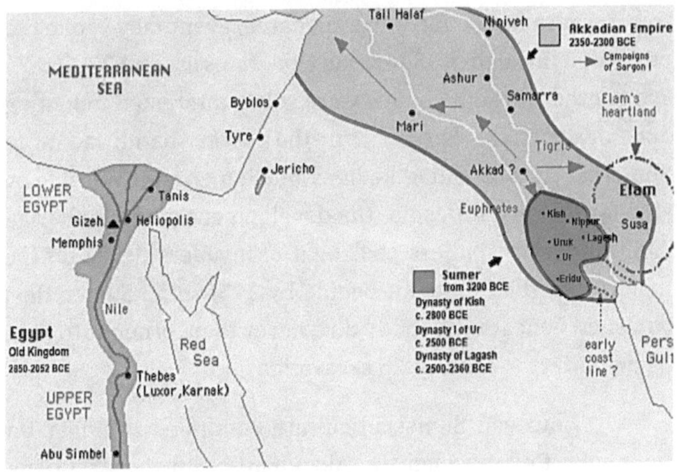

10) Map of ancient Mesopotamia and Egypt (Old Kingdom), often called the "Cradle of Civilisation."

adopting their way of life and articulating their own languages in cuneiform script.

While the Akkadian Empire had a significant surplus of agricultural products, it lacked almost everything else (particularly metal ores and timber, all of which had to be imported).[7] As a result, international trade was a top priority and developed significantly during the Akkadian period, with its influence ultimately spreading as far as the Taurus Mountains (southern Turkey), the "cedars" of Lebanon, and the copper deposits of Magan (modern-day United Arab Emirates). With a large territory of approximately 800,000 km^2 (about ten times modern-day Austria), they were able to establish control over vital imports, trade routes, and resources.

Sargon, the Akkadian monarch, eventually broke the peace and invaded the Sumerian city-states around 2300 BCE. While other invading bands were solely interested in looting and destruction, Sargon, on the other hand, admired Sumerian culture, and Akkadians and Sumerians lived side by side for centuries. Despite the development of the Sumerian civilisation, a relentless onslaught of invaders drove off the majority of the Sumerian people by 1750 BCE. Sumer then vanished beneath the sandy dirt, never to be brought to light again until 19th-century CE excavations.

However, Sumerian culture continued to linger for millennia: first through the Akkadians, then the Assyrians, and ultimately the Babylonians. Sumerian inventions, technologies and culture were passed down from the Babylonians to the Hebrew, Greek, and Roman societies. Some persist today in Iraq's contemporary marshes, the last living descendants of the Sumerians (the marsh Arabs), whose genes are approximately 71% Sumerian. It's worth noting that this ancient yet magnificent civilisation rose and fell in 2800 years, serving as a humbling reminder that the last two millennia—from 1 CE until the present—represent just two-thirds of the Sumerian odyssey!

However, only a few decades after the Sumerians' demise, ancient Egypt reached the pinnacle of its dominance in the New Kingdom (about 1500 BCE), dominating most of Nubia (Egypt's southern valley) and a major chunk of the Near East, before entering a period of steady decline. Ancient Egypt's history was characterised by a sequence of stable kingdoms interrupted by times of relative turmoil known as the "Intermediate Periods" (the Old Kingdom of the Early Bronze Age, the Middle Kingdom of the Middle Bronze Age and the New Kingdom of the Late Bronze Age).[6] By 1500 BCE, Thebes had surpassed Memphis and grown to be the world's greatest city, with a population of around 75,000 (a position it held until about 900 BCE when Nimrud overtook it).

Nonetheless, various foreign kingdoms attacked or conquered Egypt over its intriguing history, including the Hyksos, the Libyans, the Nubians, the Assyrians, the Achaemenid Persians, and the Macedonians led by Alexander the Great. The capacity of ancient Egyptian civilisation to adapt to the agricultural circumstances of the Nile River valley contributed to its prosperity. The lush valley's predictable floods and managed irrigation generated excess harvests, which supported a more dense population, as well as social development and culture.[8]

Ancient Egypt has left a lasting legacy. Their impressive works of art and cutting-edge architecture were widely imitated by the rest of the world and served as a kind of blueprint for all successive empires, influencing them in areas like practical and effective medicine, efficient irrigation systems, and agricultural production methods. In addition, they developed a system of mathematics. Therefore, they understood geometry concepts, such as determining the surface area and volume of three-dimensional shapes useful for architectural engineering, and algebra, such as the false position method and quadratic equations, which supported building the monumental pyramids.[5] All three of Giza's famed pyramids and their elaborate burial complexes were built by Pharaohs Khufu, Khafre, and Menkaure, during a frenetic construction period, from roughly 2550–2490 BCE (Remarkably, the timeframe between the construction of Giza's iconic pyramids and the rule of Gaius Julius Caesar is

11) All of the Giza Pyramids in a single photo. The Giza pyramids are among the most recognisable symbols of ancient Egyptian culture.

greater than the period between Gaius Julius Caesar is to you reading this book).

With approximately 2.3 million individual pre-formed blocks, each weighing up to 2.5 tonnes (about the weight of a pickup truck), each perfectly hand-cut and transported by boat down the Nile River in the scorching heat, elevated from ground level to eventually 139 metres, and utterly accurately constructed, the Great Pyramid of Khufu, weighing up to 5,750,000 tonnes (about 570 times heavier than the Eifel Tower) is regarded as the pinnacle of architecture and the most outstanding construction ever built. However, equally unreal is the room near the top of the Great Pyramid, also known as the King's Chamber. The King's Chamber itself is vast, with big blocks of stone used in its construction, but the ceiling is much more so, with five tiers of nine huge granite beams topped by an over-wide roof with 55 massive beams (each beam weighing about 70 tonnes, which is the equivalent of two humpback whales!). Halfway up the Great Pyramid (about 60-70 metres above ground level), engineers carefully installed the 70-tonnes perfectly pre-formed beams that had been transported by boat down the Nile River. That's a modest engineering feat for conscripted part-time farm workers using manual labour and basic hand tools.

Since each "brick" used to build the King's Chamber and the pyramid itself weighs several to tens of tonnes and is flawlessly fitted into place, it defies explanation in terms of leverage and mechanical advantage. Even today, about 4,500 years later, with modern-day technology, there is simply no satisfactory explanation as to how the Great Pyramid and its chambers were constructed. It's a baffling mystery, and it leads to debate over various theories and explanations.

But returning to our imperial journey, in addition to Egypt, Sumer, and the Indus, the only other area in the world having large cities at that point in time was what is now Peru. The Caral-Supe civilisation (also known as Norte Chico) thrived somewhere around three rivers, the Fortaleza, the Pativilca, and the Supe. Surprisingly, Caral-sophisticated Supe's culture evolved a millennium after Sumer in Mesopotamia and was contemporary with the Egyptian pyramids. The civilisation's most noteworthy achievement was its monumental architecture, which included massive earthwork platform mounds and underground circular plazas. It has been recognised as the oldest-known civilisation in the Americas. However, despite the fact that they constructed massive stone structures, they never seemed to have a writing system, pottery, or visual art.[9] But the lack of indications of conflict, such as warfare, weapons, or damaged bodies, is what makes Supe's historical legacy and rule so intriguing (much every power we've discussed thus far has faced some form of warfare).

However, returning to our imperial journey's next era, after the early Bronze Age, around 2200 BC, came the Bronze Age proper. Dividing these two periods is a very important event, known as "the 4.2 kiloyear event", which was one of the most serious climatic events of the Holocene epoch.[10] The exact cause of the event is still debated, but the most hypothesised approach is that 100 years of extreme drought may have initiated this event. When the drought brought famine to the land, no meaningful central government could respond to it, resulting in the collapse of the Egyptian Old Kingdom, the Akkadian Empire and the Indus Valley civilisation. As a result, a new set of cultures emerged.

Following the collapse of the Akkadian Empire, the people of Mesopotamia gradually merged into two large Akkadian-speaking states: Assyria in the north and Babylonia in the south. In 1894 BCE, in the southern part, Babylon's then tiny town would eventually take over the others, establishing the short-lived first Babylonian Empire, also known as the First Babylonian dynasty. The extent of the Babylonian Empire at the start and end of its reign was centred in what is now modern-day Iraq.

Until the rule of Hammurabi, its sixth Amorite king, Babylon remained a small town in a small state. Hammurabi oversaw substantial building work in Babylon, transforming it from a small, flawed town to a large metropolis befitting of royalty. Additionally, Hammurabi developed a strong bureaucracy, strict taxation, and centralised governance. After refining his kingdom, he gradually invaded southern

Mesopotamia, bringing peace to the region after turbulent periods, eventually uniting the patchwork of various rulers into a unified nation: Babylonia. Between 1770 and 1670 BCE, Babylon was thought to be the greatest city in the world, and it may have been the first to achieve a population of 150,000 humans.[11]

But moving to the northern Akkadian-speaking state, historians view ancient Assyria as the first legitimate empire, implying that they were the cradle of imperial history, thus setting the framework for every superpower that followed. At its peak in the 7th century BCE, the Assyrian Empire included modern Iraq, Syria, Lebanon, Israel, and parts of Turkey, Iran, and Egypt. Its attractions included a massive zoological and botanical park and a massive library.

However, the tale of Assyria's ascent to power began many centuries earlier, in the Late Bronze Age, in the metropolis of Ashur, a tin and textile trading hub in northern Iraq located near the Tigris River. It was named after a god who was supposed to be the city's essence and, eventually, that of the whole empire. Politics and religion were inextricably interwoven for the administration-minded Assyrians. Yet, around 1300 BCE, Ashur-uballit I, a high priest, took on the position of king and began a series of military operations, thereby transforming Assyria from a small city-state to a mighty territorial state, implying that a single centralised administrative institution was in charge of various locations, cultures, and peoples.

The empire flourished in part as a result of polishing its administrative structure and strategy of deporting and relocating local populations for various objectives, therefore cutting people's ties to their homelands and severing loyalties among local groups. When the Assyrians conquered a new region, they built towns linked by well-kept royal roadways, and when a new king took control, he would frequently construct a new capital. New palaces and temples were built and beautifully decorated with each relocation.

However, while Babylonia had been a cultural powerhouse for millennia, dating back to the dawn of writing, Assyria considered itself the successor and guardian of this legacy. After the death of Babylon's king, Hammurabi, Babylon's power weakened and was eventually conquered by the Assyrian Empire in the 7th century BCE. Yet, not long after, the Assyrian Empire experienced great problems through a succession of internal conflicts and external attacks, and its capital city of Nineveh was sacked by a coalition of Medes and Babylonians in 612 BCE. The Assyrian king, Ashur-uballit II, retreated to the city of Harran, but was finally captured and killed, and the empire collapsed. The Medes and Babylonians divided the Assyrian territories among themselves, with the Medes claiming control of the northern region and the Babylonians taking control of the south. This signified the end of the Assyrian Empire and the beginning of a new period in the Near East.

Nonetheless, the Assyrians' innovations lived onward. Their commitment to continual innovation, effective administration, and great infrastructure became the model for every empire that followed in the area and all over the world. But as for the Babylonians, they did not live happily ever after. In 539 BCE, shortly after Babylonia's reconquest, the Achaemenid Empire (Persian Empire) defeated the empire, bringing about the end of the Babylonian Empire.

But getting back on track with our great imperial journey, during Babylonian's millennia-long reign, South Asia experienced dramatic changes. The Indus Valley cities disappeared altogether, resulting in a new set of people, the Indo-Aryans, who were part of the larger Indo-European group. The group migrated from Central Asia into South Asia and was united by shared cultural norms and language. While the Indo-Aryan linguistic group is mostly found in northern India, all South Asians across the Indian subcontinent are genetic descendants of a combination of South Asian hunter-gatherers, Iranian hunter-gatherers, and Central Asian steppe pastoralists in variable proportions. But aside from Asia, it was also during the Bronze Age proper that civilisations emerged for the first time in various regions of the planet, notably Sub-Saharan Africa and, unbeknownst to many, North America. The Poverty Point culture arose in North America around 1730 BCE and is one of the oldest sophisticated cultures, likely the first tribal society in the Mississippi Delta and the modern United States.

Finally, we have Europe's first civilisation, the Minoans, on the island of Crete, whose origins date back to roughly 3500 BCE, with the sophisticated urban civilisation commencing around 2000 BCE. The Minoans constructed vast and sophisticated buildings up to four stories high, with elaborate plumbing systems and artwork. The biggest Minoan palace is that of Knossos, which had a population of 100,000 people around 1600 BCE. The Minoans' cultural influence extended beyond Crete to the Cyclades and even the Old Kingdom of Egypt through traders and artists.[12]

In contrast to every other historical entity hitherto mentioned, the Minoans met their downfall with a bang. The

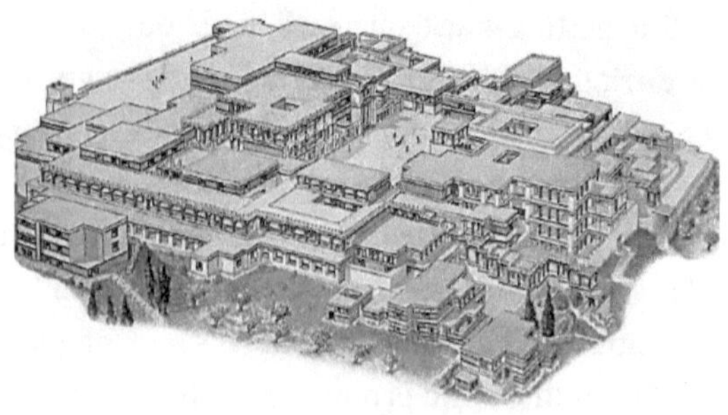

12) A reconstruction of Knossos. It became the ceremonial and political centre of the Minoan civilisation and culture.

Minoan Eruption on the island of Santorini is believed to have contributed to the fall of the Minoan civilisation in approximately 1650 BCE, making way for the Mycenaean Greeks. It was one of the greatest volcanic explosions ever observed, with a 7.0 on the (logarithmic) Volcanic Explosivity Index. The adjacent Minoan village of Akrotiri on Santorini was engulfed in a layer of ash as a result of the eruption.

In the central and eastern regions of Southeastern Europe, however, the Bronze Age began much later: approximately 1800 BCE. The Balkan area became a hub for the dissemination of mining and processing technology across Europe. Adding tin, which occurs only very infrequently in the Earth's crust, is what turned copper into bronze, a much tougher metal. Casting durable swords and other massive artefacts were made possible by the alloy bronze, and the division of labour expanded to meet the rising demand for metal products. Metallurgy, commerce, and the military's defence of resources all required a steady supply of food; thus, farming and raising livestock became a necessity rather than a luxury. Because of its inaccessibility, a new type of elite was rising, and they took charge of organising, coordinating, and protecting the metal trade on a global basis.

The development of metalworking permanently altered the look of Europe. Due to the necessity of importing some of the raw materials from far places, new networks for the exchange of products were established. As trade grew, communities from different regions began interacting with each other, transportation significantly improved, and lifestyles became more homogenous. The spread of both technical know-how and cultural customs was rapid.

The Iron Age

1200–600 BCE | Estimated World Population: 110 million

Around 1,200 BCE, a few centuries after the catastrophic Minoan Eruption, the next major imperial turning point in *Homo sapiens* history occurred: the Bronze Age collapse. For causes that are yet unknown, the world's most influential civilisations, such as Greece, Anatolia, and Egypt all vanished almost instantly. What followed is generally regarded as "the Greek Dark Ages" (a period in which the historical record for the region goes somewhat silent).

Surprisingly, many of the world's most renowned legendary tales are said to have started during this era of silence (such as those found in the Jewish Torah, the Greek Iliad, and the Hindu Mahabharata). The end of the Bronze Age coincided with the first usage of iron in the Middle East and Asia Minor. Therefore, the period in which the Greek Dark

Ages took place is also aptly called the Iron Age, the third and final period of the three-age system.

The Hittite Empire of Asia Minor is where we find the earliest evidence of iron usage, which dates back to the late second millennium BCE. The new material began its triumphant march in the first millennium BCE because it was more durable and easier accessible than bronze. After the 8th century BCE, widespread cultural exchange resulted from increased long-distance trade.

However, in addition to the Hittite Empire of Asia Minor, several centuries after the beginning of Europe's Iron Age, new cultural groupings emerged, such as in the 6th century BCE, in north-central Italy, the Etruscan civilisation, and in the 5th century BCE, the Hallstatt civilisation, which was succeeded by the La Tène culture further north in the Alps. Hallstatt benefited from the rise of the Mediterranean states to the south, mainly the Greek colonies in southern France and the Etruscans in north-central Italy, who were increasingly interested in establishing trade relationships with the peoples of central Europe. The Alpine Passes served as critical trade routes connecting the Mediterranean with Central Europe. The primary items sold south of the passes were amber from the Baltic region and salt, sometimes known as "white gold." For its flavouring and preservation properties, salt was highly prized.

Officially, the next major era in imperial history is not included in the three-age system since, beginning approximately 600 BCE, archaeology is no longer the main source of information. Nothing from the Bronze or Iron Ages remains that can be considered a history book. We have kings' names, battle lists, and other historical facts etched into rock and metal, but nothing that begins: "In this treatise, I am going to record the entire history of Sumer in exquisite detail." As a result, beginning from 600 BCE, *Homo sapiens* began meticulously analysing and writing about historical events, such as the ancient Greek historian and geographer Herodotus, nicknamed "The Father of History."

Another famous example is *The Art of War*, a Chinese military treatise written by Sun Tzu in the 6th century BCE. The 13-part treatise, with each chapter devoted to a different facet of warfare, is famous for its eternal wisdom on strategy and tactics and has impacted military thought for millennia.

But even more remarkable, it was as "the Father of History" wrote about pivotal historical events and the written knowledge of *The Art of War* was first used in practice, *Homo sapiens* entered a new era: classical antiquity.

Classical Antiquity

600 BCE–500 CE | Estimated World Population: 150 million

Classical antiquity is the period when both Greek and Roman civilisations flourished, exerting enormous influence over much of Europe, North Africa and Western Asia, laying the foundations for what would come to be known as "Western Civilisation." The classical antiquity era started with three mighty Persian empires, the Maurya & Gupta Empires in India and the first imperial dynasty in China. Nevertheless, before we look at the Persian Empire, we shall first make a little detour through China.

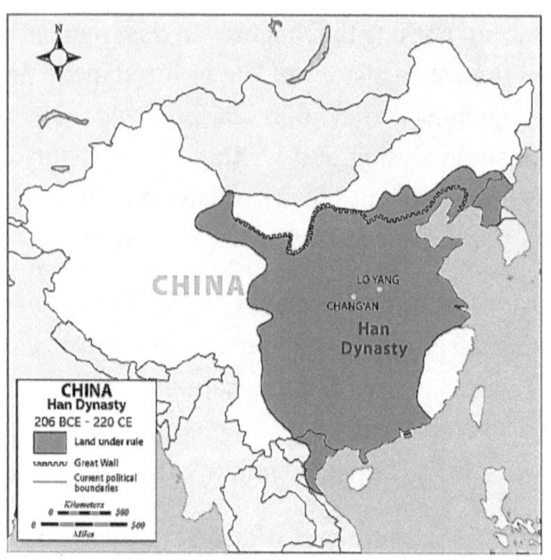

13) Map showing the extent of the Han Dynasty c. 100 BCE.

The Han dynasty (206 BCE to 220 CE) and the Three Kingdoms period (220-280 CE) were two pivotal periods in ancient Chinese history that shaped the development of Chinese culture and identity. Liu Bang established the Han dynasty, which was distinguished by its massive geographical expansion, economic progress, and cultural prosperity. During this period, the renowned Silk Road trade routes were formed, linking the Han empire to the Middle East. During the Han period, Confucianism (an ancient Chinese belief system which focuses on the importance of personal ethics and morality) emerged as the dominant political and cultural philosophy, exerting a lasting impact on Chinese thinking and values.

The Han period of 400 years of Chinese history is split into two parts. The first two centuries are known as the Western Han period (206 BCE–2 CE), as its capital was located in Chang'an in the west; the final two centuries are known as the Eastern Han period (25-220 CE) since its capital was located in Luoyang in the east. The Western Han era reached its peak about 50 BCE when it controlled approximately 6,000,000 km² of area. The pinnacle of the Eastern Han era, however, was marginally greater, as they ruled around 6,500,000 km² of land around the year 100 CE (as a comparison, both Western as Eastern eras reigned about twice the size of modern-day India).

However, near the end of the Han dynasty, corruption and political unrest brought about its final demise. This was followed by the tumultuous Three Kingdoms period, in which three nations struggled for control of China: Wei, Shu, and Wu.

Despite the brutality and disorder of the Three Kingdoms, significant cultural and technical breakthroughs arose during this time period. During this time, for instance, the classic Chinese novel *Romance of the Three Kingdoms* was written, and it is still widely read and beloved today. In addition, the discovery of paper and the widespread use of iron tools during the time of the Three Kingdoms had a lasting influence on the development of Chinese culture.

Overall, the Han dynasty and the Three Kingdoms period significantly impacted the development of Chinese history and culture, and their influence is still felt today.[13] But perhaps the most iconic remnant of ancient Chinese dynasties is the Great Wall of China. The Great Wall of China is a set of fortifications constructed along the historical northern frontiers of ancient Chinese kingdoms and Imperial China as a defence against numerous nomads from the Eurasian Steppe. It is the world's biggest man-made structure, with a total length of roughly 21,196 kilometres, including overlapping and reconstructed parts (As a comparison for this unbelievable length, if you were to attach all wall sections together, you could drive back and forth between Beijing and Paris). Some walls were constructed as early as the 7th century BCE, and the first emperor of China, Qin Shi Huang (220–206 BCE), later linked certain sections. Later on, various dynasties constructed and maintained extensive lengths of boundary walls. The Ming dynasty (1368–1644 CE) constructed the most renowned parts of the wall.

But nonetheless, continuing on our imperial trip, alongside the Chinese dynasties, the Achaemenid Empire, also known as the Persian Empire, also flourished during classical antiquity. Established by Cyrus the Great in 550 BCE, this ancient Iranian empire had an excellent infrastructure, including road and postal networks, as well as the growth of government services, including a large, professional army. The Persian Empire is well-known for instituting a successful style of centralised, bureaucratic administration, as well as for its multicultural policies, and prompted the formation of comparable governance models by a number of subsequent empires.

The Persian Empire was at its greatest territorial extent between 522–486 BCE, with a population of 17–35 million people and a land area of 5,500,000 km^2 (about ten times the size of modern-day France). However, as powerful and sophisticated as they were, not even hundreds of years later, an eager, intelligent young Greek tyrant would ruin everything for them.[14]

In barely over a decade, Alexander III, better known as Alexander the Great, king of the ancient Greek kingdom of Macedon, transformed the nature of the ancient world solely

14) The Achaemenid Empire at its greatest territorial extent, under the rule of King Darius I (522 BC–486 BCE)

by himself.[15] Alexander was born in Pella, Macedonia's ancient capital, in July 356 BCE and was schooled in his adolescence by the legendary Greek philosopher Aristotle. At the age of 20, Alexander succeeded his father, Philip II, to the throne, inherited a strong yet volatile empire, and spent the majority of his reign undertaking a protracted military campaign throughout Western Asia and Egypt.

Following his conquest of Asia Minor (modern-day Turkey) in 334 BCE, he invaded the mighty formidable Persian Empire and waged a brutal ten-year war. Against overwhelming odds, he led his army to triumph over the Persian territories of Asia Minor, Syria, and Egypt. Therefore, he is considered as one of the most outstanding, if not the greatest military commanders in history. Alexander's period in Egypt was marked by the founding of Alexandria, which eventually, in the 1st century CE, became the largest city in the world with an estimated total population of 216,000–500,000. By the age of 30, with a total surface area of

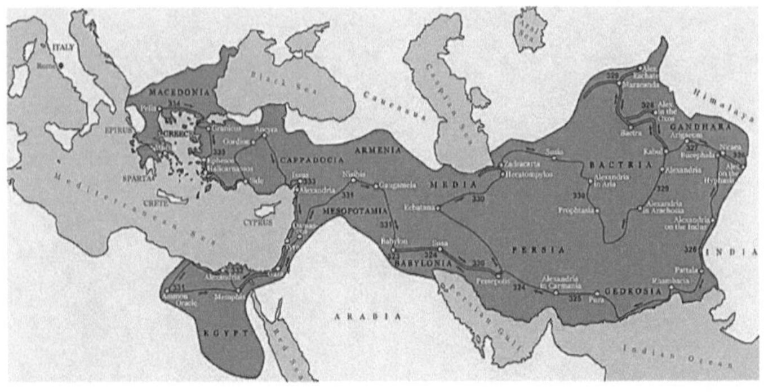

15) The Empire of Alexander the Great and his conquest course from Greece to India to Babylon (334-323 BCE).

5,200,000 km², he had created one of the largest empires in history, stretching from Greece to northwestern India.

Yet, as powerful as he was, Alexander the Great died at the age of 32 in June 323 BCE. His death was unexpected, and there are conflicting theories for what happened, the most widely accepted of which is that he died of illness. However, Alexander's death left a number of planned military and economic endeavours unfinished, beginning with a Greek invasion of Arabia. Immediately after his death, a series of civil wars raged over the Macedonian Empire, which ultimately fell to the Diadochi (the rival generals, families, and friends of Alexander the Great). Eventually, the remaining majority of the former Achaemenid Empire's land was governed by two powers: the Ptolemaic Kingdom, ruled by Ptolemy, who served with Alexander from his first campaigns and was among the seven bodyguards of Alexander; and the Seleucid Empire, ruled by Seleucus I Nicator, one of the generals of Alexander the Great after his death.

Alexander's legacy includes the cultural spread and syncretism that his conquests fostered, such as Greco-Buddhism and Hellenistic Judaism, with his death marking the beginning of the Hellenistic period. Alexander's founding of dozens of Greek cities bearing his name and the subsequent spread and amalgamation of Greek culture into the overwhelmingly influential Hellenistic civilisations extended his legacy to the far east of the Indian subcontinent—and through the Roman Empire, the Hellenistic culture (and thus Alexander's legacy) ultimately evolved into the current Western civilisation.[15]

With the history and legacy left behind by Alexander the Great, our imperial journey must go on, and we now proceed to the Romans, arguably the most important and influential imperial power in the history of *Homo sapiens*, and is rightly considered the pinnacle of ancient history. Given the importance of the Roman Empire to the modern world's foundation and the direction of the book's remaining chapters, it is worthwhile to delve more extensively into its remarkable history.

When Rome emerged into the light of history about 700 BCE, it was already inhabited by various peoples of different cultures and languages (while legend has it that on April 21, 753 BCE, Romulus and his twin brother Remus founded Rome on the place where they had been breastfed by a she-wolf as orphaned infants, the Romans themselves believed that there were descendants of refugees from the Middle East who had survived the Trojan War). Yet, even with a great fundament and various peoples of different cultures, Rome remained a small fringe tribal and italic settlement on the Tiber River for a significant amount of time.

But how did this little town eventually become one of the world's greatest empires of all time?[16] First of all, as mentioned previously, during this historical period, the Etruscans lived in the north and the large Greek colonies in the south, making Rome fortunate with its positioning (because it was positioned in between these fanatically trading groups, Rome became a crossroads of their trade routes).

Once upon a time, there were seven kings, each of whom reigned for a long time and left a valuable legacy. However, in 509 BCE, Rome was shaken by a scandal: The son of King Lucius Tarquinius Superbus, the legendary seventh king of Rome, commonly known as Tarquin the Proud, raped the Roman noblewoman Lucretia. In the end, Superbus and his royal family were expelled from Rome, leading to the Roman Republic's establishment.

Beginning in 509 BCE, the Romans vowed to never again accept such a concentration of power and instead instituted an annual election of two consuls. The consuls were controlled by the Senate, which consisted of 300 patricians (the so-called descendants of Romulus). However, those who weren't so lucky to be born into the right families joined the plebeians, also known as Plebs (meaning the crowd, the masses). Even if plebeians were as wealthy as patricians, they had no right to take any position in the state. But eventually, by 287 BCE, after much argument, the plebeians had achieved full equality of rights. The unity of Rome found its best expression in the formula "Senatus Populus Que Romanus" (SPQR), meaning "the Senate and People of Rome."

In 390 BCE, Rome's history could have ended when the city was unexpectedly attacked by the Gauls (a group of

Celtic peoples from the La Tène culture in the Alps). At the battle of Allia (390 BCE), the Roman army was annihilated by a confederacy of Celtic tribes led by the Senones and Brennus. The Celts went on to lay waste to Rome before being routed by the exiled Roman general Camillus.

After this brutal clash with the Gauls, the shaken Romans slowly recovered and conducted a military reform, drastically transforming and dividing the Roman army into Manipula (meaning a handful), making it more mobile in battle. From 390–280 BCE, Rome waged constant wars, eventually conquering much of Italy. But instead of imposing a hefty tax on the conquerors, the Romans imposed an alliance treaty in which the loyal allies provided Rome with an endless stream of recruits. This structure enabled the Roman legions to hold their ground against the most efficient fighting force of the time: the all-conquering Macedonian phalanx, led by Pyrrhus (a relative of Alexander the Great). After the five years protracted Pyrrhic War (280–275 BCE) and having conquered the Greek city of Tarentum, southern Italy, Rome now faced a far more dangerous adversary: Carthage (modern-day Tunisia), also known as the lord of the Mediterranean.

In 146 BCE, after fierce battles between Rome and Carthaginians, the capital, Carthago (whose eventual peak population of 500,000 made it the second-largest city in the western Roman Empire), was finally conquered. The city was wiped out, and its population was enslaved, allegedly ploughing salt into the air as an eternal curse. The Romans eventually lavishly rebuilt a new city on the same land, ultimately becoming one of the most prosperous cities in the Mediterranean.

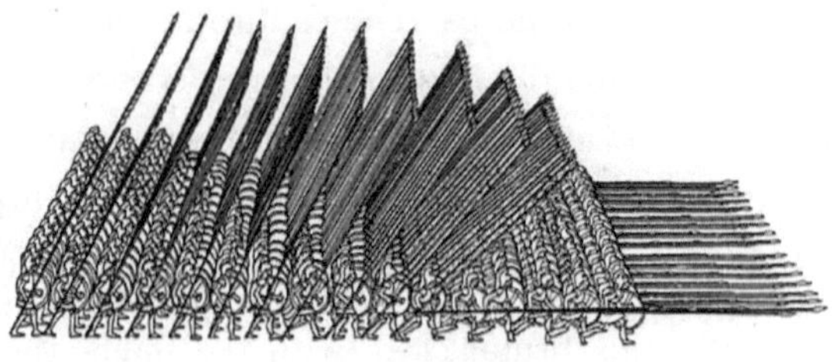

16) An illustration of the Macedonian phalanx. This one is formed of 256 men.

Additionally, in 146 BCE, the Romans destroyed another city, Corinth (close to Athens), making Greece and Macedonia Roman provinces. Rome usurped the colossal wealth of Alexander the Great's disintegrating empire, but Rome's patriarchal simplicity succumbed to sophisticated Greek culture: Greek culture effectively became a second-state language. Interestingly, while ancient Greece remained a collection of squabbling city-states, Rome gradually extended citizenship rights to the conquerors, laying down the basis for an empire. This was absolutely a total game-changer. Not very long after this, in 133 BCE, the city of Rome was the first ever city in *Homo sapiens* history to reach the incredible milestone of a population of 1,000,000 (to fully comprehend how remarkable this landmark is, consider the landmarks of previously discussed cities: According to Ian Moris,[2] around 32 generations before Rome, Nineveh, the Assyrian capital, was the first to attain a population of 100,000).

However, returning to its history, having conquered great territories, Rome fell victim to globalisation, as the inflow of unpaid slave labour bankrupted the small farmers. Previously united, the Senate and the Roman people split into two hostile camps (civil wars gripped Rome). Amid this chaos, Gaius Marius, a popular general, rose to power, offered social mobility for the proletariat, and began enrolling the proletariat into the army with a promise of a land grant at the end of service, making the legions personally devoted to their generals.

In 49 BCE, after things had started to calm down, two outstanding generals fought over Rome. Gnaeus Pompeius, who had conquered the eastern provinces for Rome (from southeast Turkey to Jerusalem), cleared the Mediterranean of piracy and defeated Spartacus's slave revolt. The other famously outstanding general, Gaius Julius Caesar, who had conquered Gaul (encompassing present-day France, Belgium, the Netherlands, Luxembourg, and parts of Switzerland, Germany, and Northern Italy), defeated significant southwest parts of Germania (Central Europe and Scandinavia) and invaded Great Britain.

According to Roman law, a general had to dismiss his legions before returning to Rome, to have in return a moment of glory: a triumphal entry into the capital to the applause of the citizens. However, Caesar performed a hitherto unseen manoeuvre: he refused to submit to the Senate and marched his legions to Rome. It would take him several years to defeat Gnaeus Pompeius and his other rivals, putting Roman legions against each other. In the process, Caesar annexed new territories and gave Cleopatra the Egyptian throne. After a romantic cruise along the Nile, Cleopatra gave birth to Caesarion, also known as "Little Caesar." On his return, Caesar added "Emperor" to his name (originally meaning victorious commander) and became a dictator by gaining control of all political positions (from Consul to tribune of the plebs).

Rumours spread that Caesar wanted to declare himself king, and conspiracy was brewing in the Senate. Sooner or later, in 44 BCE, Julius Caesar, the Roman dictator, was assassinated by a group of senators (Collectively, the group stabbed Caesar a reported 23 times).

Caesar left his wealth unexpectedly for all concerned to his grandnephew, 19-year-old Gaius Octavius, also known as Octavian or Caesar Augustus. Octavian immediately joined in the power struggle and defeated his last rival, the general and politician Marcus Antonius in 31 BCE, who likewise had an affair with Cleopatra (the lovers paid the ultimate price: with their lives). After this, Octavian was left as ruler of a vast territory.

While Julius Caesar ruled for four years, Octavian, who gave himself the title "Augustus" (meaning the venerable or the great rule), ruled for an endless 43 years. He didn't "formally" abolish the republic; he simply took control of all possible positions making his power almost absolute, and modestly called himself "Princeps" (meaning first senator). And even though skirmishes with barbarians continued along the borders inside them, the Empire experienced a great economic upswing and a period of "Pax Romana" (Roman peace) and stability, which was set to last 200 years.

After Augustus' power became hereditary, the senatorial opposition left us with vivid portraits of the first emperors. In the first century CE, most emperors reigned relatively shortly (after Augustus, between 14 and 98 CE, there were as many as 11 emperors, including Aulus Vitellius, who was the emperor for only eight months). The second century CE would go down as the era of the good emperors. In

17) The Roman Empire at its greatest extent.

117 CE, under the emperor Trajan, with a land area of 5,000,000 km² (ten times modern-day Spain), the Empire reached its largest extent. Rome connected its new territories via a network of extremely advanced paved roads (which still determines the transport map of today's Europe!). Around this time, in the middle of classical antiquity, a very crucial thing happened: a trade route opened up between the mighty thriving empires between the Western parts of Eurasia and the Eastern parts of Eurasia. This trading route, famously known as the "Silk Road", connected the Western parts of Eurasia and the Eastern parts of Eurasia for the first time. It was a primary conduit for efficient trade between the Roman Empire and China and later between medieval European kingdoms and China. The exchange of information gave rise to new technologies and innovations that would change the world.

However, returning to the emperors, it was under Marcus Aurelius, son of Commodus, that the Pax Romana ended as he preferred his gladiator's glory over the affairs of the state (the Roman state found Commodus's gladiatorial combats to be scandalous and disgraceful). Conspirators had the emperor strangled by a fellow fighter, the slave Narcissus. Upon his death, the Senate declared him a public enemy and statues of Commodus were demolished, and Rome sank into chaos.

In 212 CE, emperor Caracalla, half North African and half Syrian, granted Roman citizenship to nearly all free citizens of the Empire. The idea that you could be a true Roman in Africa or any other corner of the Empire might well be the main legacy of Rome still in use today.

However, by the middle of the century, Rome was already in the midst of such a crisis that the entire provinces began to split off, after which Roman emperor Diocletian divided the empire into four, with four co-rulers and four capitals closer to the frontier. Rome lost its significance, the Senate became a city council, and the country was now ruled by an army of officials who reported personally to the emperor. The old world centred around the concept of a free community came to an end.

After Diocletian's departure, the co-rulers were fighting for power, and eventually, in 306 CE, Constantine, the future Saint Constantine the Great, emerged victorious (he also became the first Roman emperor to convert to Christianity). After more than 300 years of persecution, Christians were now entitled to build churches alongside the

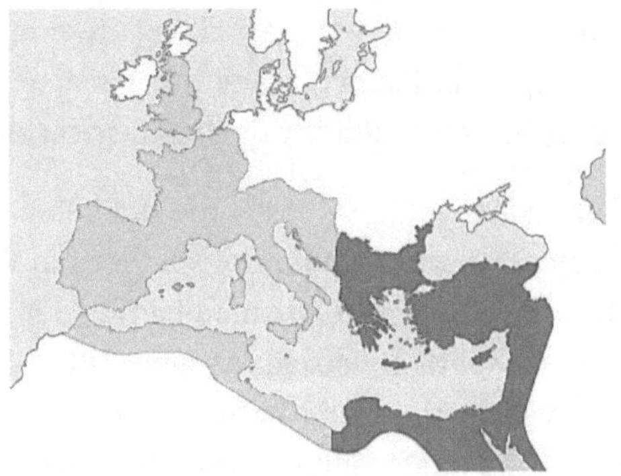

18) The division of the Empire after the death of Theodosius I

temples of Augustus and Mars. Constantine would take the cross from Jerusalem to the new capital of the Roman Empire: Constantinople (modern-day Istanbul, Turkey).

Theodosius, who was Roman Emperor from 379 to 395 CE, was instrumental in establishing the Nicene Creed as the doctrine for Christianity and started to destroy the ancient temples. Theodosius was the last emperor to rule the entire Roman Empire before the administration was permanently divided among his sons between the two separate empires (one in the west, the other in the east). The eastern half, known as Byzantium, would live another thousand years. The western part (simply known as the Western Roman Empire) however, would fall victim to the great migration of people. Rome, founded by migrants, would fall to the onslaught of a new wave of refugees. Ironically the last ruler of Rome would be called Romulus (the name of the mythic founder of Rome).

After the mighty pillars of the Western Roman Empire fell, the West fell into turmoil. But moving away from

Europe, in addition to the partial collapse of the most influential empire in history, many other major events and powers occurred concomitantly during classical antiquity.

For example, there was the Kingdom of Aksum (modern-day northern Ethiopia and southern Arabia), which was established when the long-standing Kingdom of Kush in ancient Egypt's south collapsed. Aksum (100–940 CE) was deeply involved in the trade network between the Indian subcontinent and the Mediterranean (the Roman Empire, later Byzantium), exporting ivory, tortoiseshell, gold, and emeralds and importing silk and spices. While the famed economically important Silk Road was far beyond their borders, they benefited from a major transformation of the maritime trading system that linked the Roman Empire and India. With a land area of 1,250,000 km² at its zenith (about four times modern-day Italy), the Kingdom of Aksum is considered one of the ancient world's four great powers during classical antiquity.

Finally, when we zoom out of our imperial adventure, we discover that during classical antiquity, another influential civilisation evolved and thrived in a very remote region of the world. About 1400 BCE, a group of individuals settled on the lush floodplains of the Gulf Coast region and immediately discovered how to cultivate crops such as maize and beans, and their population rose fast: the Olmec culture.

The Olmec culture, a pre-Columbian civilisation, is renowned for its art and architecture, which included giant stone heads that continue to confound researchers. They also established a sophisticated religious system in which gods such as the jaguar and the feathered serpent played a significant role. When the Olmec culture grew, it began to

19) Tikal Temple V is one of the major pyramids at Tikal, the capital of a conquest state that became one of the most powerful kingdoms of the ancient Maya.

impact the surrounding Maya culture and is, therefore, considered the "mother culture" of Mesoamerica. Much of the Olmec's religious beliefs and creative traditions were absorbed by the Maya, who dwelt further south in what is now modern-day Mexico, Guatemala, Belize, and Honduras.

Preclassic (2000 BCE–250 CE), Classic (250–900 CE), and Postclassic (950–1539 CE) are the three key periods that comprise the intriguing Maya civilisation's history. These were preceded by the Archaic Period (8000–2000 BCE), during which the first settled villages and early agricultural developments emerged. Yet, the Maya did not just borrow from the Olmec. They created their own distinct complex culture, complete with a complicated political structure, with strong kings and queens ruling over city-states, a writing system, intricate mathematics, the renowned Mayan Calendar, and spectacular technical achievements such as the construction of enormous pyramids and cities, including the ancient capital city Tikal, Uaxactún, Copán, Bonampak, Dos Pilas, Calakmul, Palenque, and Río Bec. Yet, regarding the pyramids, at the site of El Mirador, Guatemala, is the pyramid La Danta, which is, with an estimated volume of 2.8 million cubic meters, the largest pyramid in the world (The Great Pyramid of Giza is roughly 2.6 million cubic metres).

Nevertheless, while the Mayans reached their zenith in the middle of the Classic Period (with a population of ten million), the Classic Maya civilisation finally crumbled and abandoned their cities by the end of the 9th century CE (with the last dated monuments in the region matching to 889 CE), for reasons that historians continue to argue, including overcrowding, environmental deterioration, conflict, shifting trade routes, prolonged drought, and uncontrollable self-poisoning of the freshwater to curb unsustainable population growth. The collapse of the Classic Maya is one of archaeology's biggest unsolved mysteries. (It's possible that a multitude of reasons contributed to the collapse).

But getting back on track, the vanishing of the fascinating, mysterious and unique Maya civilisation marks the end point of our imperial adventure through classical antiquity. Nonetheless, aside from all the hitherto discussed empires, it was even more important that humankind's current moral and philosophical foundations were laid during classical antiquity. Simultaneously yet independently, we got Greek philosophy, the Bible, the Hindu and Buddhist scriptures and the writings of Confucius. We also saw the first experiments with democracy and the first evidence of truly scientific thinking. But this golden age of human development did not last forever, as it came to an end around the year 500 CE with the fall of the Western Roman Empire.

The Middle Ages

500–1500 CE | Estimated World Population: 250 million

The Middle Ages, sometimes known as the mediaeval period, was a period of enormous cultural growth and change, highlighted by the fall of the Western Roman Empire, the establishment of the Catholic Church, the beginning of feudalism, and the transition to the Renaissance and the Age of Discovery. This era saw the construction of magnificent cathedrals and castles, the development of towns and commerce, and the flourishing of the arts and sciences. From tales of knights in shining armour to the devastating Black Death, the Middle Ages is replete with narratives that continue to inspire and intrigue us to this today. Let's explore the most important people, events, and ideas that shaped the world as we know it today as we travel through this remarkable historical period.[17]

Population collapse, counter-urbanisation, the collapse of centralised power, invasions, and mass migrations of groups, which began in late antiquity, persisted throughout the early Middle Ages. The Migration Period's large-scale movements, which included numerous Germanic peoples, founded new kingdoms in what remained of the Western Roman Empire. In some locations, the classical period ended earlier (as in China) or later (as in Mesoamerica), but most of the main classical civilisations eventually disintegrated across the world. Although it is commonly used, the name "Dark Ages" doesn't really correspond to this period partly because the so-called "dark" ages were only restricted to Western Europe, as other areas during this time, such as the Middle East and China, were, in fact, experiencing new golden ages.

So what made things in Western Europe change so drastically? Well, numerous theories have been put forward, but one thing we know for certain is that it involved large-scale migrations in several parts of the world. Whether or not those migrations were sparked by climate change or some other natural phenomenon, we can't be sure. But in the end, it ended up creating a rebellious domino effect.

It all started when the Huns (nomadic people who lived in Central Asia) moved into Europe and pushed several Germanic tribes south, bringing them into conflict with Rome. They forced them into Roman lands, undercut Rome's tax base, and demanded expensive tribute. When they were gone, they left pure destruction and chaos behind. Eventually, those Germanic people, known to the Romans as "barbarians", caused the Western Empire's fall and plunged Europe into the so-called Dark Ages.

But it wasn't simply barbarians who wreaked havoc. A major eruption in early 536 CE (or maybe late 535 CE) blasted enormous amounts of sulphate aerosols into the sky, reducing solar radiation reaching the Earth's surface and cooling the atmosphere for many years. The 536 CE volcanic winter was the most severe and prolonged period of climate cooling in the Northern Hemisphere in the last 2,000 years. According to mediaeval scholar Michael McCormick, 536 CE was the worst year in history: "It was the beginning of one of the worst periods to be alive, if not the worst year."[17]

Additionally, not long after that, there followed the First Bubonic Plague, also known as the Justinian Plague (541–549 CE). The plague ravaged the whole Mediterranean Basin, Europe, and the Near East, wreaking havoc on the Sasanian Empire, the Byzantine Empire, and notably

Constantinople. Several historians believe the First Bubonic Plague was one of the worst pandemics in history, killing an estimated 15-100 million people over the course of two centuries, equating to 25-60% of Europe's population at the time of the first outbreak.

While the volcanic winter, accompanied by the Plague of Justinian, caused tragic crop failures and famine, it didn't initiate the fall of the Eastern half of the Roman Empire. Yet, it did set the stage for the rise of a new great and influential civilisation nearby: Islam.

Born in Mecca in the year 570 CE, Muhammad ibn Abdullah, also known as Muhammad the Prophet (in Arabic مُحَمَّد), was raised in a respected but poor family and grew up to be a successful merchant. At the age of 40, Mohammed had a profound religious experience and began receiving revelations from Allah (in Arabic الله)that were later recorded in the holy book of Islam, the Quran. He began to preach the message of monotheism and soon gained a following, but he also faced opposition from the ruling elites in Mecca.

In 622 CE, Mohammed and his followers fled Mecca for Medina in a journey known as the Hijra, marking the start of the Islamic calendar. In Medina, Mohammed established a community of believers and continued to preach the message of Islam. Over time, he gained more followers and military strength and eventually conquered Mecca in 630 CE, leading to the spread of Islam throughout Arabia. Mohammed continued to receive revelations from Allah and led his community until his death in 632 CE. He is considered by Muslims to be the final prophet in a line of prophets that

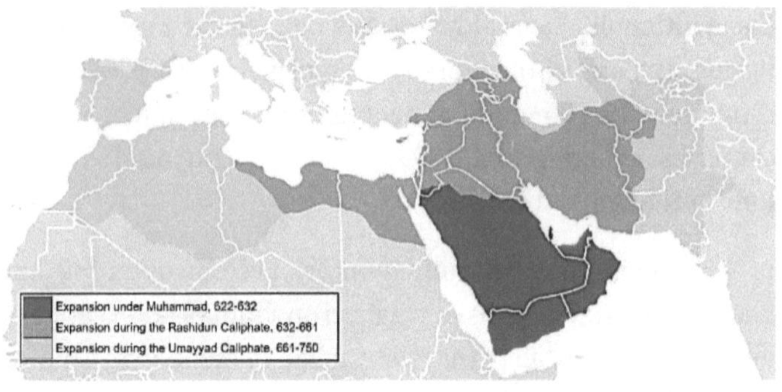

20) Map showing the expansion of the Islamic Caliphates in the 7th and 8th centuries CE.

includes Adam, Abraham, Moses, and Jesus, and is revered as the messenger of Allah and the model for Muslim behaviour.

After the death of the Prophet Muhammad, a series of four major dynasties (Islamic Caliphates) that ruled the Islamic world followed. The first caliphate was the Rashidun Caliphate, which lasted from 632 to 661 CE and was characterised by a period of rapid expansion and political stability.

The second caliphate was the Umayyad Caliphate, which ruled from 661 to 750 CE and saw the capital move from Medina to Damascus. The Umayyads built a vast empire and developed a sophisticated administration system. It was the Umayyad Caliphate during which Islam reached its greatest territorial peak, encompassing 13,400,000 km² of land (about 24 times modern-day France, or slightly larger than the Roman and Persian Empires combined!), making it the world's largest empire yet seen and the sixth-largest ever to exist in history. Its borders stretched from the Iberian Peninsula (Spain and Portugal) to as far as Pakistan.[18]

The Abbasid Caliphate (named after the uncle of the Prophet Muhammad) was the third caliphate and eventually overthrew the Umayyad caliphate and ruled as the Abbasid caliphate until 1258 CE, when the Mongols invaded the capital city of Baghdad, bringing an end to the Abbasid period of cultural renaissance and prosperity and so the Islamic Golden Age. But before its demise, political ties were established with the courts of the Chinese, Byzantine, and Charlemagne emperors, ancient writings were translated, and lasting achievements were made in geography, philosophy, medicine, astronomy, and mathematics, ushering in an era of cultural and scientific prosperity.

The fourth and final caliphate was the Ottoman Caliphate, which ruled from 1299 to 1924 CE and controlled a large empire that spanned three continents. Despite its military and political power, the Ottoman Caliphate faced internal decline and was eventually dissolved after World War I. These four major caliphates played a crucial role in shaping the Islamic world and had a lasting impact on the political, cultural, and religious landscape of the region.

But as Islam, as both faith and power, grew from the Middle East to its nearby neighbours, Christianity expanded in Europe, whereby Germanic tribes were converted (or re-converted from Arianism) by missionaries of the Catholic Church. Even outside the Roman Empire, Goths (Germanic people) converted to Christianity. Eventually, areas that started out as barbarian kingdoms evolved into major powers such as England, France, and the Holy Roman Empire.

But there were many other things happening in the world during the Middle Ages. In Southern Africa, for instance, there emerged the metropolis of Great Zimbabwe, which included the first ever gigantic constructions in that region. Also, in West Africa, we get several major empires for the first time, including the Mali Empire (1214–1255 CE) with Mansa Musa as its ruler, who is considered the richest person ever in all of world history (though it can be difficult to fairly calculate a fortune based on gold, salt and land, some estimates put Mansa's modern-day net worth at around $400 billion!).

When Mansa Musa gained control in 1312 CE, most of Europe was in the grip of famine, poverty and civil war. However, many African kingdoms and the Islamic world were thriving, and Mansa Musa was the key to bringing the fruits of that prosperity to his own domain.

The Mali Empire originated as a minor Mandinka kingdom on the upper banks of the Niger River and later acquired access to trans-Saharan trade routes with the conquest of Sosso (1235 CE). Mansa Musa established control over even more important trade routes between the Mediterranean and the West African Coast by deliberately annexing Timbuktu and reestablishing sovereignty over Gao, prolonging a phase of growth that greatly extended Mali's size. What was also very advantageous was that its territory was rich in natural resources such as gold and salt.

Under King Mansa Musa, the empire became the most fabled power in mediaeval Africa, with schools and mosques in hundreds of densely inhabited cities. The king's wealthy heritage was passed down through generations, and mausoleums, libraries, and mosques still exist as a reminder of Mali's golden period. Unfortunately, it would be the Mandinka themselves who would bring the empire to its demise. Mahmud Keita IV, the final king of the Mali Empire, died about 1610 CE. According to oral legend, he had three sons who battled for Manden's remains. Following the death of Mahmud Keita IV, which marked the end of the Mali Empire, no single Keita ever governed Manden.

But despite the fact that we have just covered the Mali Empire, the fall of the Western Roman Empire, the establishment of the Catholic Church, and the advent of the mighty Islam, there was one imperial power that was much bigger than those we have already discussed: the Mongol Empire.

The Mongol Empire is considered the biggest land empire of all time, reaching from Korea to Ukraine and Siberia to southern China. However, before the Mongol Empire was established on the vast plains in the 12th century CE, the East Asian steppe was home to scattered bands of Mongol and Turkic pastoral nomads led by Khans. These people travelled between summer and winter campsites while tending to their flocks of sheep, cattle, yaks, and camels. But as these nomadic tribes regularly engaged in conflict, Temujin, a member of an aristocratic Mongol family, was destined to change that. Despite growing up in poverty and losing his father at a young age, he rapidly ascended to power by forming clever partnerships and alliances with other leaders. Temujin promoted troops on the basis of merit and equally divided the rewards among them. Yet, his most ingenious tactic was dispersing the vanquished nomads among his own warriors to prevent them from banding together to revolt. These innovations made him unstoppable, and by 1206 CE, Genghis Khan, often known as the First Great Khan Emperor, unified the inhabitants of the felt-walled tents.[19]

Northern China and the eastern Islamic countries were first conquered by the Mongols under Genghis Khan. His family, known as the Golden Lineage, received the Divine Mandate after his death in 1227 CE. The sons and daughters of Genghis Khan sacked the Russian princes and the Turks of Central Asia in the 1230s, and in 1241 CE, they conquered Eastern Europe by invading Russia and Bulgaria. By the 1250s, the Mongols had conquered the Islamic lands all the way to Baghdad, and by 1279 CE, their influence had spread to southern China. During these campaigns of Genghis Khan, Kublai Khan, and Timur between 1206 and 1405 CE, between 20 and 57 million people were killed as a result of battles, sieges, early biological warfare, and massacres. (This accounts for roughly 10% of the world population at that time and does not include deaths from the Black Death in Europe, West Asia, or China).

Around this time, the Mongols encompassed about 23,000,000 km^2 of territory—about 16.11% of the entire world land area—and became the largest contiguous land empire in world history (the British Empire, which was established about a century after the Mongols broke up, had a substantially larger size, but its territories were scattered across the globe). Let that substantial number sink in for a moment: its (peak) surface area is similar to 1.35 times modern-day Russia, nearly 2.5 times modern-day United States, or just slightly larger than the (peak) territory of the Han dynasty, Persian, (Alexander's) Greek, and Roman Empires combined!

However, life within the Mongol Empire was more than simply war, looting, and destruction. Once a province was captured, the Mongols left their internal politics alone and relied on local authorities to rule it. The Mongols tolerated all religions as long as the leaders prayed for them.

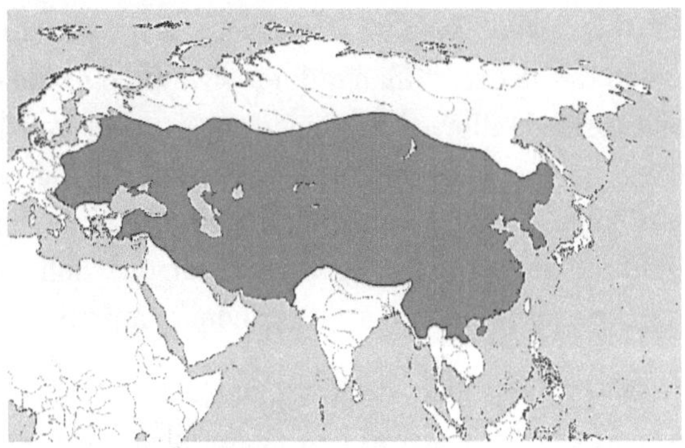

21) A map showing the Mongol Empire at its greatest extent.

Although they often abducted craftsmen, academics, and engineers, they recognised their value and forcefully relocated them across Asia to continue their work.[19]

Trade developed along the Silk Road as much of Eurasia became politically united. At sea, trade was thriving, particularly in blue-and-white porcelain, which mixed white pottery from Mongol China with blue dye from Mongol Iran. The Mongols prized Chinese gunpowder technologists in particular—Chinese monks discovered in the 9th century CE that the mixture of saltpetre (potassium nitrate), sulfur, and charcoal that became known as "gunpowder" could burn and explode as a propellant.

Incidentally, it was along the Silk Road that the famed Marco Polo (1254–1324 CE), a Venetian merchant, travelled across Asia during the Mongol Empire's peak, and recorded his travels in the book *The Marvels of the World and Il Milione*, which outlined (to Europeans) the then-mysterious culture and inner workings of the great wealthy Eastern world.

This great wealth, however, was not to last. Succession to the Great Khan did not always go to the oldest son but instead allowed siblings, uncles, and cousins to compete for leadership, with older widows serving as regents for their sons. By the 1260s, Genghis Khan's grandchildren were embroiled in a full-fledged civil war over inheritance, dividing the country into four independent empires.

The struggle between Batu and Jochi, Genghis Khan's second and third sons, who had acquired the western and central parts of the empire, was the most crucial. Rivalries over territory, resources, and power drove the decades-long struggle. Despite internal tensions, the Mongol Empire conquered additional lands, notably in Europe and the Middle East. However, the civil conflict undermined the empire's control over these distant lands and led to its partition into Khanates. The Mongol civil war affected the empire and the world, as it reduced Mongol authority over acquired lands and empowered local authorities to contest Mongol rule. Infighting and instability hampered trade routes and slowed cultural interchange, isolating most of the Mongol realm. Yet, despite these hurdles, the legacy of the Mongol Empire and its impact on world history is still felt in the modern world.

However, if we return to the path of our imperial voyage, soon after the fall of the mighty Mongol Empire, other great empires emerged, including those in America. Great cities existed in pre-Columbian North America, contrary to popular belief. Cahokia existed in what is now Illinois, and there were also Puebloan cities in New Mexico. Hundreds of Ancestral Puebloan dwellings can be discovered across the American Southwest. These Puebloan towns and communities, virtually all of which were built before 1492 CE, can be found all across the Southwest. Cahokia was the largest pre-Columbian Native American city (between 1050-1350 CE) and the most important urban settlement of the Mississippian culture, which developed advanced societies across much of what is now the Central and Southeastern United States beginning more than 1,000 years before European contact. The Cahokia Mounds are now regarded as the largest and most sophisticated archaeological site north of Mexico's great pre-Columbian cities.

However, apart from any Middle Age superpowers mentioned thus far, it was around the end of the Middle Ages that one of the most important occurrences in the history of *Homo sapiens* took place: Early in the morning of October 12, 1492 CE, after more than two months of arduous travel, Italian adventurer Christopher Columbus made it ashore with his ship Santa Maria in what is now the Bahamas, known by its local people as Guanahani. He later explored the well-known islands of Cuba and Hispaniola, establishing a colony in what is now Haiti. In early 1493, Columbus returned to Castile, carrying with him a number of captive indigenous. The news of his revolutionary trip quickly spread throughout Europe. Columbus claimed in his letter on his first journey (1492-1493 CE), written after his first return to Spain, that he had reached Asia, as previously recorded by Marco Polo and other Europeans. Following this successful and historic voyage, Columbus went on to make a second (1493-1496 CE), third (1498-1500 CE), and fourth trip (1502-1504 CE), paving the way for broad European exploration and colonisation of the Americas.[20]

Interestingly, Columbus was one of the last explorers to reach America, not the first. A courageous party of Vikings

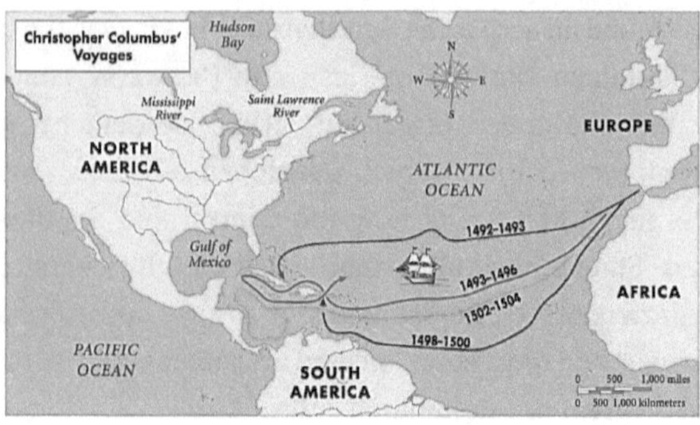

22) The routes of the four trips of Christopher Columbus.

led by Leif Eriksson set foot in North America and founded a colony around 500 years before Columbus.

However, returning to the Americas, in addition to these pre-Columbian Native American cities and settlements in the north, in Mesoamerica, the mighty Aztec Empire, and in South America, the great Inca Empire reigned.

The Aztec Empire, also known as the Triple Alliance, comprised three Nahua city-states: Mexico-Tenochtitlan, Tetzcoco, and Tlacopan. These three city-states governed the territory in and around the Valley of Mexico from 1428–1521 CE, when they were conquered by a joint force of Spanish conquistadores and local allies commanded by Hernán Cortés. The Aztec Empire was the last of the great Mesoamerican empires to fall before the arrival of the Europeans. They constructed magnificent temples, employed complex agricultural techniques, and offered human sacrifices to their gods.[21] The Spanish conquest of the Aztec Empire was a critical part of the Spanish colonisation of the Americas. After the empire fell, its temples were defiled or destroyed, and its exquisite art was melted down and used to make coins. Ordinary people were afflicted by European-introduced illnesses that killed out up to 50% of the population, and their new masters were no better than the Aztecs.

But, just as important as the Aztecs, the Spanish arrived at the Inca Empire's borders in 1528 CE, further south. The Inca Empire was a brilliant and enthralling society that concluded thousands of years of Andean culture. The Inca Empire was preceded in the Andes by two large-scale empires: the Tiwanaku (300-1100 CE), centred around Lake Titicaca, and the Wari or Huari (600-1100 CE), centred around the city of Ayacucho. Scholars consider the Andean culture to be one of only five pristine civilisations in the world. Under the command of Sapa Inca (supreme leader) Pachacuti-Cusi Yupanqui, whose name meant "earth-shaker," they began a far-reaching expansion in 1438 CE. After conquering the Tribe of Chancas (now Apurmac), he was given the name Pachacuti. During his reign, he and his son Tupac Yupanqui conquered most of modern-day Peruvian territory.[22]

Pachacuti reorganised Cusco's empire into the Tahuantinsuyu, consisting of a central government led by the Inca and four powerful regional governments controlled by powerful chiefs. The famous Machu Picchu is supposed to have been constructed by Pachacuti as a family house or summer retreat (however, it might also have been an agricultural outpost). Remarkably, the Incas constructed the temple in perfect harmony with celestial and planetary movements. It was a holy temple where sacrifices and religious rites were performed. Because it was sacred, only priests and high-ranking Incas were permitted to enter.

Traditionally, the army was controlled by the Inca ruler's son. Túpac Inca Yupanqui, Pachacuti's son, began conquests to the north in 1463 CE and continued them as Inca rulers following Pachacuti's death in 1471 CE. The Kingdom of Chimor, the Inca's sole serious rival for the Peruvian coast, was Túpac Inca's most important conquest. Túpac Inca's empire then spanned modern-day Ecuador and Colombia.[22]

The Inca Empire was remarkable in that it lacked several characteristics associated with Old World civilisations. According to anthropologist Gordon McEwan, "the Incas were able to build one of the greatest imperial powers in human history without the use of the wheel, draught animals, knowledge of iron or steel, or even a writing system." The Inca Empire operated mostly without a monetary system or marketplaces. Instead, products and services were exchanged based on reciprocity between people as well as between individuals, tribes, and Inca kings. But yet, even without these technologies, in less than a century, the Inca Empire quadrupled its territory from around 400,000 km^2 (1448 CE) to roughly 1,800,000 km^2 (1528 CE). This enormous expanse of territory had a wide range of cultures and climates—scholars believe that the Inca Empire had a population of more than 16 million at its zenith.

However, just several years later, between 1532–1572 CE, led by Francisco Pizarro, explorer and conquistador, the Spaniards touched the Inca border and held violent campaigns that ultimately resulted in the conquest of the Inca Empire.

On November 16, 1532 CE, in one of his most renowned tactical military tactics, Pizarro devised a trap for

23) Depiction by John Everett Millais (1845 CE) of Pizarro seizing the Inca Emperor Atahualpa.

Inca emperor Atahualpa. Pizarro lured Atahualpa to a feast in honour of the Emperor with fewer than 200 troops against several thousand. Atahualpa agreed to send roughly 5,000 unarmed troops to the feast. At the sign, the Spaniards opened fire on the unarmed Incas. As the panicked unarmed Incan warriors were trapped in close quarters, they became easy prey for the Spanish. In less than an hour, Pizarro's soldiers slaughtered all 5,000 Incans.[23]

Regarding the gruesome Spanish violence, the deadliest weapons during the Spanish conquest of the Inca Empire were not physical weapons but, just as during the conquest of the Aztec Empire, diseases: between 1533–1572 CE, diseases ravaged the Inca people. Ultimately, there were around 7,700,000 native deaths from the probable typhus, influenza, smallpox, and measles outbreaks (Ever since the end of the last Ice Age, about the start of the Agricultural Revolution, when the landbridge Bering Strait got flooded, *Homo sapiens* living in the Americas got separated from the rest of the world, including their diseases, and therefore lacked immunity to them).

Pizarro succeeded in infiltrating the Inca Empire's heart by taking advantage of the conquistadors' weakening of the Inca Empire, the death of Crown Prince Ninan Cuyochi, and the resulting civil war between the brothers Atahualpa and Huáscar. With their royalty and centre of worship decimated, the ordinary Inca people quickly accepted Spanish control, resulting in local cooperation that allowed the Spanish to entirely overrun the area by 1572 CE, thereby ending the Inca Empire.

After the downfall of the mighty Inca Empire, history arrives at the Late Middle Ages, nearly transitioning to the sixth and final era: the Modern Age. One could argue that the transition started with the most deadly pandemic of all time: the Black Death (also known as the Second Bubonic Plague). It is the most lethal epidemic in recorded human history, killing 75–200 million people and peaked in Europe between 1347-1351 CE (in many European cities, it killed up to 50% of the population). Bubonic plague is caused by the bacterium *Yersinia pestis*, which is carried by fleas, but it may also be transferred through person-to-person contact via aerosols, resulting in septicaemic or pneumonic plagues.

Yet, others claim that the Late Middle Ages ended with the collapse of the Byzantine Empire (Eastern Roman Empire) and the fall of its magnificent capital Constantinople. But how could this millennium-old city, which in large part represented the pinnacle of urbanisation for human history, eventually tumble?

It was all caused by the mighty Ottomans, who successfully breached Constantinople's ancient land wall. Osman I founded the Ottoman Empire in the late 13th century and built a minor province, or principality, in northwestern Anatolia in the town of Söğüt (modern-day Bilecik Province, Turkey). Interestingly, this tiny province outmanoeuvred stronger neighbours and ended up establishing a massive Ottoman Empire in just a few generations.

The Anatolian peninsula was a patchwork of Turkic states wedged between a disintegrating Byzantine Empire and a weakening Sultanate of the Seljuk of Rum during Osman's rule. Osman swiftly extended this region with these neighbours by a combination of shrewd political alliances and military conflicts, drawing mercenaries first with the prospect of treasure, then later with his reputation for victory. Osman was the first in a long series of Ottoman kings known for their political acumen: they increased their power by fighting alongside particular factions when necessary and against them when the moment was right, frequently putting political and military utility over ethnic or religious affinity.

Following his father's death, Osman's son Orhan developed a sophisticated military structure and tax-collecting system aimed at supporting rapid territorial expansion. The Ottomans' first significant expansion occurred in the Balkans, in southeast Europe. The military was made up of Turkic fighters as well as Byzantine and other Balkan Christian converts. By the end of the 14th century CE, not long after refining its military, the Ottomans invaded or subjugated the majority of Anatolian beyliks as well as the Balkans.

However, as Sultan Beyazit I concentrated on Western development, the Central Asian king Timur launched an invasion from the east. He kidnapped Beyazit and imprisoned him, sparking a ten-year war that almost annihilated the Ottoman Empire. Sultan Murad II eventually won this war but fell short of one of his most ambitious goals: conquering Constantinople, the Byzantine capital.

Sultan Mehmed II, popularly known as Mehmed the Conqueror, was determined to succeed where his father had failed. In preparation for the attack on Constantinople, he recruited a Hungarian engineer to manufacture the world's biggest cannon, employed Serbian miners to dig tunnels beneath the city's defences, and instructed his fleet of warships to be moved overland, approaching the city from an unexpected direction. After besieging the city for 55 days, Constantinople eventually fell to the Ottomans in the spring of 1453 CE. It would become the Ottoman capital, known by its common Greek name, Istanbul, meaning "to the city." However, Constantinople was just a glimpse of its former glory by the time Mehmed II conquered it. Yet, it thrived once again under Ottoman rule. The Ottomans would continue to grow, consolidating their political power and profitable trading routes. Ultimately, at the end of the 17th century CE, the Ottoman Empire consisted of an area of 5,200,000 km^2 (slightly larger than the Roman Empire), extending from Hungary to the Persian Gulf and from the Horn of Africa to the Crimean Peninsula.

However, with the rise of the Ottomans and having meandered through dozens of great empires, from the Sumerians, Romans, and Mongols to the Incas and Ottomans, our imperial journey has come to an end. Nevertheless, as history evolved from one crossroads to the next, significant scientific discoveries were made, paving the way for the "immortality" of humankind. And as we shall shortly see in the next chapters, around 1500 CE, history made its most pivotal decision, one that changed not just the fate of *Homo sapiens* but, potentially, the fate of all life on Earth: the Modern Revolution.

PART IV

THE MODERN REVOLUTION

0) As mission Commander Neil Armstrong, the first man on the Moon, stated on July 20, 1969: "That's one small step for a man, one giant leap for mankind."

The Formation of the Modern World

11. THE SCIENTIFIC REVOLUTION

According to some scholars, the beginning of the Modern Age may be traced back to the 19th century, with the advent of the Industrial Revolution. Yet, in the opinion of others, it all started in 1789 CE with the French Revolution. Some people date it even further back, to when Europeans first began colonising the Americas in the 16th and 17th centuries, or towards the end of the Renaissance period, with the 1543 CE Nicolaus Copernicus publication *De revolutionibus orbium coelestium* (also known as *On the Revolutions of the Heavenly Spheres*) often cited as its beginning.[3] However, from the perspective of Big History, the above events are only accelerations of a rapid expansion of humankind's collective knowledge, an expansion that began with marginal steps in East Africa around 200,000 years ago.[1] Yet, it's ridiculous to label Lucy or il principe, who foraged as "early-early-early modern." So, to simplify things for the remaining chapters, we'll assume that the early modern period started about 1500 CE (with the understanding that everything is a bit arbitrary).

Nevertheless, as we shall shortly see, in the last 500 years, the power of *Homo sapiens* has skyrocketed. In 1500 CE, there were about 500 million *Homo sapiens* on Earth; today, there are more than eight billion! In 1500 CE, only a handful of cities had populations of over 100,000 people, and "skyscrapers" were often no more than three stories tall and made of materials like mud, wood, and straw. There were people, horses, goats, chickens, and carts all over the muddy streets, and it was either extremely hot or chilly. In the city, you could always hear people or animals, as well as tools like hammers and saws. A candle or torch was all that illuminated the city after dark. For virtually all its history, humanity was confined to Earth. Birds and gods all belonged in the sky. But on July 20, 1969, humans successfully landed on the Moon. This was the greatest cosmic and evolutionary triumph of humanity.[0]

The Age of Science

Nonetheless, reaching this milestone was anything but straightforward. We can only trace the roots of contemporary science back to the 16th century CE, which was also a time of vast exploration, religious and political upheaval, and amazing literary creation. Leonardo da Vinci, a Renaissance painter and innovator, started his famous *Mona Lisa* in 1503 CE, and it was completed three years later (The Renaissance was a fervent period of European cultural, artistic, political and economic "rebirth" following the Middle Ages).[3] There were other developments in natural history, geography, cosmography, and mathematics. Engineering, mining, navigation, and the military arts all saw significant innovation in this century.

Classical antiquity's principles and accomplishments gave this "rebirth" new life to surpass them. One may trace the origins of the first sparks of astronomical science and ideas on the tangible workings of the earthly universe all the way back to classical antiquity. In his work *On the Heavens*, written in 350 BCE, the Greek philosopher Aristotle presented three compelling reasons supporting the theory that the Earth is spherical rather than a flat plate.[2] To begin, he understood that lunar eclipses occur when Earth moves in between the Sun and the Moon. Only if Earth was spherical would its shadow always appear round on the Moon. Unless the eclipse always happened when the Sun was directly under the centre of a flat disc representing Earth, the shadow would have been extended and elliptical. Second, the Greeks observed that the North Star seemed lower in the southern sky than it did in the northern sky throughout their travels. Since the North Star is located above the North Pole, it will seem straight overhead to an observer at the North Pole, but it will appear near the horizon to someone gazing at it from the Equator. The sails of a ship may be seen coming over the horizon before the hull, which was a third argument the Greeks used to conclude the Earth is round.

The Sun, the Moon, the planets, and the stars, in Aristotle's view, revolved around a stationary Earth. This was due to his mystical conviction that Earth was at the centre of the cosmos and his preference for circular rather than linear motion. Ptolemy, writing in the 2nd century CE, developed this concept into a whole cosmological paradigm. The Earth was in the centre, with the Moon, the Sun, the stars, and the five known planets at the time—Mercury, Venus, Mars, Jupiter, and Saturn—revolving around it on eight spheres.[2]

The celestial body locations may be predicted with a fair amount of precision using Ptolemy's model. However, Ptolemy assumed the Moon followed a course that brought it twice as close to Earth as at other times so that he could accurately anticipate these places. Therefore, there should be periods when the Moon seems twice as large as usual.

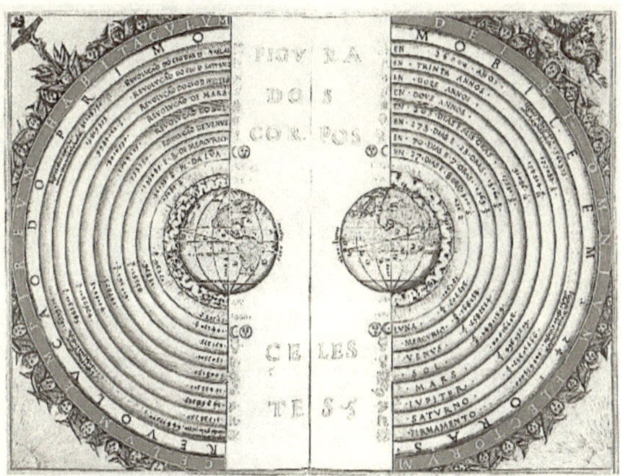

1) A depiction of the Ptolemaic Universe as described in the *Planetary Hypotheses* by Bartolomeu Velho (1568 CE).

Ptolemy was aware of this shortcoming in his model, yet it did not prevent widespread acceptance of his theory. It had the huge benefit of leaving masses of space beyond the sphere of fixed stars for heaven and hell. Thus, it was chosen by the Christian church as the depiction of the world that was consistent with Scripture.

However, going back to the Scientific Revolution, in 1514 CE, a Polish monk named Nicholas Copernicus presented a more straightforward alternative (Copernicus initially shared his model under cover of anonymity, perhaps out of concern that the church would label him a heretic). His theory was that the Sun remained fixed in the sky while the Earth and the planets revolved around it. A century and a half elapsed before anyone truly considered his theory. Yet, the German astronomer Johannes Kepler and the Italian astronomer Galileo Galilei were the first to openly embrace the Copernican hypothesis, despite the fact that its projected

2) Galileo showing the Doge of Venice (the chief magistrate and leader of the Republic of Venice) how to use the telescope.

borbits did not exactly match the ones seen. In 1608 CE, the Aristotelian and Ptolemaic cosmological axioms were dealt a fatal blow. As of that year, Galileo began using the newly developed telescope to study the night sky and noticed that Jupiter was accompanied in its orbit by a number of smaller satellites or moons.

Homo sapiens owe much of our progress to the development of the telescope. Several persons have laid claim to the invention of the telescope, but it was the Dutch lensmaker Hans Lippershey who is generally credited with the invention in 1608 CE. He supposedly came up with the concept for his invention after observing two children in his shop using two lenses to create the optical illusion of a nearby weather vane. In 1609 CE, however, Galileo built his own telescope after hearing about the perspective glass and directed it at the stars.

From the vastness of space to nature's microscopic complexity, both a microscope and a telescope may be used to magnify things such that they can be seen by the human eye. Microbes make up 99.99% of all life on Earth, but for the vast majority of Earth's history, humans knew very little about them. In 1674 CE, Dutch microbiology pioneer Anton van Leeuwenhoek gazed into a drop of pond water with his handmade microscope and saw a universe of tiny organisms swimming around. He saw beneath the microscope "joyfully swimming animalcules," to use his words. After making this discovery, he shares it with experts, but they dismiss him as crazy. "We suspect you might have drank too much gin before you wrote this letter," they allegedly said. Finding a whole new world living in a little puddle of pond water? To some extent, it's understandable that they didn't immediately believe him. Just because someone claims something doesn't mean it is true without further scientific investigation. The origin of microscopy occurred when experts were dispatched to the Netherlands to confirm his significant findings.

The Scientific Revolution is the period of time in history that paved the way for van Leeuwenhoek, Alamogordo, and the Moon landing. As a result of this revolution, *Homo sapiens* has acquired tremendous new powers as a result of its investment in scientific research. The emphasis on abstract reasoning, quantitative thought, an understanding of how nature works, the view of nature as a machine, and the development of an experimental scientific method marked the beginning of the end for the Greek view of nature that had dominated science for nearly 2,000 years prior to the Scientific Revolution.

Over the course of the past five centuries, humans have come to believe that they may improve their abilities by funding scientific inquiry. It wasn't based on wishful thinking; it has been objectively verified time and time again.[3] The wealthier classes and governments were more likely to invest in scientific research when there was more evidence to back it up. Without these funds, we would never have been able to accomplish many great things, such as the marvels of going to the Moon, genetically modifying species, or splitting the atom. But why did humanity start to think they might acquire superpowers through science? What acted as the glue that permanently joined science, government, and business?

Since the Cognitive Revolution, at the very least, *Homo sapiens* have been curious about our place and purpose in the Universe. Our forefathers invested a lot of energy into the quest to understand the laws that govern the natural world. But there are three major ways in which contemporary scientific knowledge departs from all other bodies of knowledge: acceptance of one's own lack of knowledge, the importance of observation and precise mathematical analysis, and the development of previously unknown abilities. Interestingly, despite popular belief, the Scientific Revolution did not usher in a new era of enlightenment. It has been a revolution driven by ignorance, first and foremost. The realisation that humankind does not have all the answers was the epiphany that sparked the Scientific Revolution.

The ability to accept a lack of understanding has allowed modern science to flourish, making it more adaptable, inquisitive, and dynamic than any other body of knowledge.[0] Our ability to comprehend the Universe and develop novel technology has been greatly bolstered as a result. Interestingly, modern society has shown a higher tolerance for ignorance than at any other time in history. The widespread adoption of an almost religious faith in technology and the techniques of scientific study, which have partially replaced the belief in absolute truths, has contributed to the stability of modern social systems.

Empirical observations have been gathered throughout history, although their impact has typically been negligible. When we already have all the information, why spend time and money gathering more data? However, contemporary society has admitted that it lacks knowledge on many crucial topics, prompting a search for entirely new knowledge. That's the goal, and scientists are working hard to make it happen.

The Mathematical Principles of Natural Philosophy, written by Isaac Newton in 1687, is widely regarded as the most seminal work in the Modern Era. Newton put out a universal theory of movement and change. The tremendous achievement of Newton's theory was to describe and predict the motions of all bodies in the Universe, from falling apples to shooting stars, using just three mathematical laws:

◊   *A body remains at rest, or in motion at a constant speed in a straight line, unless acted upon by a force.*

◊   *When a body is acted upon by a force, the time rate of change of its momentum equals the force.*

◊   *If two bodies exert forces on each other, these forces have the same magnitude but opposite directions.*

With these principles, Isaac Newton proved that the laws of nature are mathematical in nature and is therefore considered one of the most influential scientists of all time, laying the foundations of classical mechanics and becoming the cornerstone of modern physics.

The next breakthroughs in physics, the Theory of Relativity (formulated by Albert Einstein) and Quantum Theory (established by Niels Bohr and Max Planck), were only prompted by a handful of discoveries that did not fit well with Newton's rules, and they were only discovered near the beginning of the 20th century.

The Costly Gamble

The mathematical language of modern science is difficult for the human mind to understand, and its conclusions sometimes contradict common sense, making it difficult for the general public to fully digest it. Despite this, science has achieved an extremely high level of reputation as a result of the new capabilities it has granted humanity. Even if they aren't familiar with the technicalities of nuclear physics, presidents and generals have a solid understanding of the destructive potential of nuclear weapons.[0]

Science has given us a lot of useful tools over the years. Some are only mental, while technological tools are becoming increasingly vital. People today sometimes conflate science with technology because of how closely they are linked. We tend to believe that there is little value in conducting scientific studies if it does not lead to the development of new technologies and that it is impossible to produce such technologies without such research. However, science is not only tied to offensive mechanisms, as it is also crucial to human defences. Many progressive nations now hold the view that technical measures, rather than political ones, are the best way to counter terrorism.

Tanks, atomic bombs, and fighter jets are only a few examples of the surprising recent rise in interest in military weaponry. Changes in military organisation, rather than advances in technology, sparked the great majority of revolutions up to the 19th century. Of course, as we saw in the previous chapter, technological gaps played a significant influence in several historical encounters. However, even in those situations, few people considered expanding or creating technological gaps on purpose. Most empires didn't grow because of scientific geniuses, and their leaders didn't give technological advancement any concern.

Homo sapiens began to believe that true development was conceivable once contemporary civilisation confessed that there were many fundamental things that it still didn't know, and once that admission of ignorance was coupled with the concept that scientific discoveries might offer us new capacities. Many people started to believe that all problems could be solved if only humans had the time and resources to learn enough about them. Humanity was not destined to suffer from eternal poverty, disease, conflict, starvation, old age, and death. Those outcomes were only the result of our lack of knowledge.

However, a lot of money has to be invested in scientific research in order to learn about the human immune system, archaeology, physics, and many other great disciplines. The commitment of governments, industries, institutions, and individual benefactors to put vast sums of money into the scientific study during the past 500 years has allowed contemporary science to accomplish incredible feats. The primary reason political, economic, or religious organisations support scientific research is the hope that it will contribute to the success of such organisations' missions.

No amount of brainpower would have helped if enough resources weren't made accessible. For instance, if Darwin never existed, we would now credit Alfred Russel Wallace, who independently proposed the notion of evolution by natural selection only a few years after the birth of Darwin's theory of evolution. In contrast, neither Darwin nor Wallace would have obtained the empirical evidence they needed to construct the theory of evolution if European powers had not supported geographical and biological studies across the planet. It's quite unlikely that neither Darwin nor Wallace would have ever attempted it.

The rise and success of European empires may be attributed to more than just scientific progress. One crucial element, "capitalism", lies behind the rapid growth of both science and empire. Over the past 500 years, the concept of this crucial element has swayed the general populace to have an ever-greater faith in the future. Because of this faith, loans were established, which in turn stimulated economic growth, which in turn bolstered faith in the economy's long-term viability and paved the path for even more loans to be extended. Before this, in premodern history, there was little credit, thus less growth, resulting in little trust in the future. The explorations of Christopher Columbus and Neil Armstrong would not have happened if they weren't funded by capitalists hoping to achieve a profit.

The Rise of Capitalism

Economic history's fundamental importance is not simple to understand. It has been written at length on how the pursuit of wealth has led to the birth and destruction of nations, the discovery of new worlds and the enslavement of millions, the advancement of technology and the extermination of countless species. However, only one term is truly necessary to comprehend contemporary economic history: growth.[4] The economy of the modern era has been hyperactively expanding, devouring everything in sight and rapidly getting bigger. However, the scale of the economy has been relatively stable for a long time. Global output did rise, but that was mostly because of people and their need to possess more space. Also, the total output per person did not change.[0] But how is this possible? What is the reason for our exponential growth?

It has all to do with banks. Since banks may lend out $10 for every $1 they physically own (in their savings), 90% of the money in circulation is not backed by any form of currency other than paper coins and bills. Banks are vulnerable to a mass withdrawal of funds because of the institution's high dependence on depositors (unless the government steps in to save it).[4] It sounds suspiciously like a massive Ponzi scheme, a type of fraud in which new investors are used to paying off older ones. But, as Harari states, "if it turns out to be a fraud, then the entire global economy is a scam".[0] It's not a trick; it's a praise of the incredible powers of the conscious imagination of *Homo sapiens*. Our faith in the long-term success of our economy is what keeps banks alive (much of the world's currency relies solely on this faith). As we have seen, money is very remarkable since it can serve for so many things and be exchanged for practically any other thing.

For a very long time, humanity was doomed to repeat this fateful cycle. This kept economies from freezing completely. Only in our modern times, with the advent of a new system built on confidence in the future, have we found the key to breaking through this trap. It was an arrangement whereby participants agreed to use "credit" as a substitute for actual currency when purchasing "imaginary goods." The system is based on the idea that in the future, we will have access to substantially more resources than we have now. If we can create things now using money that will be pouring in later, many intriguing possibilities will open up. It wasn't that nobody understood the concept or couldn't figure out how to implement it. The challenge was, however, that few people wanted to take on significant debt since they didn't believe the future would be substantially better. Therefore, they were both at a disadvantage: People had a hard time getting loans to start enterprises since credit was restricted; Because of a lack of new companies, economic expansion was hampered; Because of its lack of expansion, investors and lenders were very cautious. So, as expected, progress slowed to a halt.

Over the course of history, the concept of progress has swayed the public to have a steadily increasing amount of faith in the future. Because of this faith, credit was established, which in turn stimulated economic expansion, which in turn bolstered faith in the economy's long-term viability and paved the path for even more credit. The economy wasn't like a balloon and didn't start rising suddenly. Nonetheless, the long-term trend remained clear, even after taking into account the inevitable setbacks. Because of the existing availability of credit, both governments and businesses may readily get massive, long-term loans at reasonable interest rates that significantly exceed their ability to repay.[4]

The first and most important tenet of the new capitalist religion is that "the gains of output must be spent in growing output." This is the very reason behind capitalism's name. It is important to note that "capital" in a capitalist economy is distinguished from "wealth" in a more general sense. Capital includes monetary assets, physical assets, and human resources that are used to create value. Meanwhile, money is lost on useless pursuits or forgotten about. The argument that "production earnings must be reinvested in even more output" is too simplistic to warrant serious consideration. However, most people throughout history saw it as strange.

In its original form, capitalism was a concept regarding how the economy worked. It did double duty as a description of the monetary system and a prescription for achieving rapid economic expansion by reinvesting corporate earnings back into the business.[4] However, capitalism evolved into something far more than merely an economic concept. It now includes a code of conduct, or a set of beliefs on how society as a whole should behave.[0] Public and private sectors frequently share the financial burden of supporting the scientific enquiry. Questions such as: "Can this initiative enable us to improve output and net profit?" Or "Is it going to boost the economy?"

Powerful conquerors or governmental and military leaders, such as the Ottomans, founded the majority of the early modern non-European kingdoms. They paid for their battles by taxation and looting (without drawing clear boundaries between the two), owed nothing to credit institutions, and gave even less consideration to the needs of financiers.[0]

Both the rise of modern science and the development of European imperialism were critically reliant on the capitalist economic system. Capitalist debt as we know it today was originally created because of European imperialism. In the 16th century, Spain became the most powerful European power thanks to this structure, which allowed it to exert influence over a wide global empire. It dominated a large swath of Europe, large portions of the Americas, the Philippines, and a line of outposts along the shores of Africa and Asia. The harbours of Seville and Cadiz annually welcomed vessels loaded down with goods from the Americas and Asia.

Around this time, the Netherlands was just a tiny, blustery marsh with no significant natural resources that belonged to the Spanish monarchy. However, within only 80 years after declaring their independence from Spain, the Dutch trumped the Spanish and their Portuguese allied forces as the rulers of the ocean routes, built a global Dutch empire and became Europe's wealthiest empire. The Dutch success factor was a well-developed credit system. The Dutch nobles, who did not really enjoy land battles, contracted with mercenary soldiers to defend them against the Spanish. Simultaneously, the Dutch were returning to the sea in large fleets of their own. Military explorations, especially those involving foreign mercenaries and cannon-wielding fleets, were indeed extremely expensive. However, the Dutch were able to fund theirs more easily than the powerful Spanish Empire since they earned the trust of the developing

European financial system at a time when the Spanish king was recklessly undermining that confidence. Investors provided the Dutch money to build armies and ships, which allowed them to dominate global trade routes and reap enormous financial rewards. The loan repayments facilitated by the income bolstered the confidence of the investors. Not only was Amsterdam one of Europe's busiest harbours, but it was also becoming the financial powerhouse of the world.

Therefore, the Dutch Empire was constructed by Dutch merchants rather than the Dutch government. The Spanish monarch persisted in attempting to pay for its wars by increasing taxes, despite widespread disapproval. The Dutch businessmen borrowed money to fund their conquests, and progressively they also sold shares in their enterprises that gave the buyers a stake in the businesses' future revenues. While investors were wary of giving money to the monarch of Spain, they were more confident in investing in flourishing Dutch companies.

While the bang of the costly Spanish cannons continued to be heard close to Amsterdam's walls in 1602, the most renowned Dutch cooperative firm, the Vereenigde Oostindische Compagnie (VOC) or in English, the United East India Company, was founded.[5]

With the capital it acquired from the acquisition of stock, the VOC constructed boats, sailed these to Asia, and returned merchandise from China, India, and Indonesia. The corporation used the money to arm its warships and fight off rival businesses and pirates. The inevitable invasion of Indonesia was funded by money from the VOC. As the VOC sailed the Indian Ocean, the Dutch West Indies Company (WIC) sailed the Atlantic. The WIC established New

3) Return of the second Asia expedition of Jacob van Neck (Dutch navel officer) in 1599 CE, one of the Dutch forays into the East Indies spice trade that led to the establishment of the Dutch East India Company (by painter Cornelis Vroom).

Amsterdam on an island at the mouth of the Hudson River to establish control over river traffic. Both the Indians and the British launched many attacks on the settlement before taking it in 1664. Once known as New Amsterdam, the British renamed it New York. Wall Street, one of the most renowned roads in the world, was established on top of the remnants of a wall constructed by WIC to protect its settlement from Indians and the British.

Towards the close of the 17th century, the Dutch lost New York and its position as Europe's economic and colonial powerhouse due to carelessness and expensive Atlantic conflicts. France and Britain were in a heated competition for the open position. France first appeared to be the more strong alternative since it had a larger and more experienced military, greater wealth, and a larger population than Britain. Despite this, unlike France, Britain successfully gained the trust of the banking industry.[5]

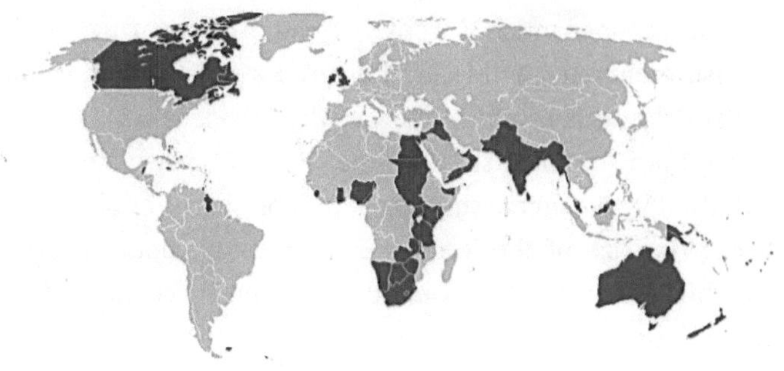

4) The British Empire at its greatest territorial peak in 1921 CE.

Loans were readily available to the British at cheap interest rates, whereas France had trouble getting loans and paid higher interest rates. The French monarch borrowed progressively more sums at ever-increasing interest rates to keep up with the mounting debts. In the 1780s, after succeeding to the throne upon his grandfather's death, Louis XVI discovered that paying interest on his debts was eating up about half of the yearly budget and that France was going to be flat broke! To find a way to resolve the dilemma, Louis XVI unwillingly called a meeting of the French parliament, the Estates General, in 1789, resulting in the French Revolution.

At the same time as France's foreign kingdom was collapsing, the British Empire was swiftly growing. A similar model to the earlier Dutch Empire, the British Empire was founded and managed by commercial joint-stock firms listed on the London stock market. Also, the British East India Company's mercenary army was responsible for conquering the Indian subcontinent rather than the British government. Even compared to the VOC, this business did exceptionally well. It maintained a strong military force of up to half a million troops, much surpassing the British army of the British monarchy, and it administered a great Indian colony for roughly a century. India and the business's huge army were both nationalised by the British crown in 1858. Eventually, encompassing 35.5 million km² (about seven times the size of the Roman Empire, or in modern terms, roughly four times the United States or twice the size of Russia), they became the largest empire in history.

European Imperialism

From the point of view of Big History, the explorations of Europe deserve recognition in the episode titled "Modern Revolution" since they finally unified all four world zones into a single global system. The possibility for widespread knowledge was greatly enhanced by the growth of an interconnected web of inventors. Yet, Europe did not become a centre of significant military, political, economic, or cultural advancements until the late 15th century. What, if anything, pushed Europeans to advance at such a tremendous speed? Interestingly, there are a fair amount of reasons.[1]

One, Ottoman control of overland trade links with Asia, notably after the fall of Constantinople in 1453, pushed Europeans to seek alternate routes to the population and wealthy regions of the East[1] (The notion that the Ottoman empire blocked Europe's eastern trade routes is a popular historical fallacy. However, the rise of the Ottoman Empire dramatically increased the cost of European trade. Due to the existence of a single power in the Middle East, the Ottoman Empire monopolised land trade from the East).

As a result of his gross underestimate of the globe's circumference, Columbus wrongly assumed that Asia was far closer than it actually was, making his journey look doomed to failure. But fortunately for him, between Europe and Asia existed a whole other continent (the Americas) that he had never heard of before. Unsuccessful finding funding in Portugal, Columbus moved to Spain. In the eyes of many, Christopher Columbus's plan to cross the Atlantic Ocean and reach Asia seemed absurdly implausible. However, he vowed to his patrons, Spain's King Ferdinand and Queen Isabella, that he would lead an expedition to Asia and return with gold, spices, and silks. However, since Spain was preoccupied with the Reconquista (the Christian reconquest of Spain and Portugal) and put all of its resources into this war, Queen Isabella initially disregarded Christopher Columbus' request for finance for a transatlantic expedition. During the course of the Reconquista, the Spanish crown used force to drive out all Muslims and Jews from Spanish territory. Yet, after seven years of requesting, Queen Isabella finally consented to fund Columbus' voyage after Spain's victory in the Reconquista.

When Columbus set sail in 1492 from Palos de la Frontera, Spain, it took him a gruelling two months and nine days to make the trip. Columbus thought he had arrived at a little island off the coast of East Asia, and because he mistook the locals for the Indies, the natives were nicknamed "Indians." For the rest of his life, Columbus continued to believe this falsehood. For him and many others of his generation, the notion that he had discovered a previously uncharted continent was beyond the realm of possibility. Ever since the beginning of the Agricultural Revolution, when the Bering Strait thawed away, making it impossible for *Homo sapiens* to cross between Asia and the Americas, only Europe,

Africa, and Asia were present to anybody in the Old World, not just the best philosophers and academics.

It was only several years after Columbus died from age-related causes on May 20, 1506, in Valladolid, Spain, that writings contended that the new territories found by Columbus were not islands off the coast of East Asia but rather a whole continent previously unknown to classical geographers and modern Europeans. A distinguished German mapmaker called Martin Waldseemüller was persuaded by these arguments and published an updated world map in 1507. This map was the first to depict the area where Europe's westward-sailing ships had landed as a distinct continent. After Waldseemüller drew it, he had to think of a name for it. However, the name "America" was given by Waldseemüller, who mistakenly thought that the continent had been discovered by Amerigo Vespucci, an Italian navigator who led many voyages to the Americas between 1499 and 1504. Many other mapmakers reproduced the Waldseemüller map, increasing the dissemination of the name he had given to the new continent.

After this seemingly minor but in hindsight major error in the Waldseemüller map, more accurate maps were created, notably the world-famous Salviati World Map. In 1525, Lorenzo Salvaiti, an ardent traveller and cartographer, desired to create the most accurate map of the world to date. Years were spent by Lorenzo obtaining information from sailors, merchants, and other travellers. He investigated old writings and the works of other cartographers to develop a map that mirrored the then-current understanding of the world. After years of work, the Salviati World Map was finally

5) The Salviati Planisphere world map. It is believed to have been drawn by Nuño García de Toreno, the head of the Casa de la Contratación (The House of Trade).

complete. It was a stunning piece of art with exquisite intricacies and vivid hues. The map depicts the world as it was understood at the time, with Europe, Africa, and Asia showed in exquisite detail, and the vast uncharted regions of the Americas and the Pacific Ocean hinted at with a feeling of mystique. Word immediately spread about the Salviati World Map, and it soon became one of the most coveted treasures in all of Europe. Kings and queens, nobility, and merchants all desired a copy of the map since it was rumoured to hold the potential to reveal the mysteries of the world. As years passed, the Salviati World Map became a symbol of travel and discovery, motivating innumerable trips and missions to map the world's uncharted regions. And to this day, the map remains one of the greatest artefacts of the age of exploration, a tribute to the brilliance and perseverance of the instinctive desire of *Homo sapiens* to explore the unknown.

However, coming back, the second factor that pushed Europeans forward so quickly was the way their culture was organised and adopted by others.[1] Many technological gadgets, such as steam engines, were not unknown to the Muslims or Chinese (which could be freely copied or bought through trade). However, they lacked the values, mythology, judicial system, and sociopolitical capitalist institutions that the West had developed over many centuries and which were not easily transferable. Because they already shared the most significant British beliefs and social systems, France and the United States moved fast to emulate Britain. Different ways of thinking and organising society meant that the Chinese and Muslims lagged behind. Thus, for the Europeans, establishing a scientific discipline was an imperial project, and constructing an empire was a scientific endeavour.

Europeans have a long history of adopting a rational, market-based way of life before their technical prowess really began to shine. In the earliest days of the technological explosion, Europeans were the best at making use of the opportunities it presented. The common mentality between a scientist and a colonists naval officer was a crucial component in cementing the historical relationship between modern science and European imperialism. Both the scientist and the conqueror first admitted their lack of knowledge by saying, "I don't know what's out there." For example, in 1798, when Napoleon conquered Egypt, he brought along 165 academics with him who were both driven to see the world and learn new things. And with this newfound information, they expected to rule the world. These empires, which owe much of their strength and influence to their collaboration with scientific endeavours, may defy easy categorisation as either good or evil (Their sins would fill one encyclopaedia, and their accomplishments would fill another). Yet, they made it possible for us to live in the world we do, including the beliefs we use to evaluate them.

If only the Aztecs and Incas had taken a more active interest in their surroundings and realised how European culture and values were structured and what the Spaniards had done to their neighbours, they could have fought the Spanish invasion more vehemently and successfully.[0] The indigenous Americans weren't the only ones who paid a high price for being narrow-minded. As soon as word of Europe's groundbreaking discovery reached Asia, everybody—including the Ottoman and Chinese dynasties—took notice. Nonetheless, they didn't seem to care much about these European findings and kept on thinking that everything happened in Asia, so they didn't try to challenge the Europeans for the Americas or the new sea routes between the Atlantic and the Pacific. While smaller European countries like Scotland and Denmark undertook exploration and colonisation missions to the Americas, neither the Islamic world nor India or China ever dispatched an expedition to the Americas.

For three hundred years, Europeans had sway over both the Atlantic and Pacific oceans, including all of North America. Significant conflicts in those areas were exclusively between European powers. The Europeans were able to invade Africa, overthrow its powers, and then split the continent among themselves because of the money and resources they had amassed. It was too late by the time the Ottomans, Indians, and Chinese finally started paying interest. (Non-European civilisations did not develop a fully global perspective until the 20th century).

The third reason that drove Europeans to develop so rapidly was that European states were relatively tiny in comparison to some of the enormous Asian empires (a vast region with a greater number of different kingdoms, such as Europe, may be more prone to conflict than were the density of various kingdoms is less significant, such as Asia). Because of this greater density, European powers were in continual competition and conflicts with one another for resources (In 1775, 80% of the world economy was based in Asia).[1] In general, conflicts during the last 500 years accelerated technical progress as technologies were adapted to meet the unique challenges faced by the military.

And finally, the fourth factor that pushed Europeans to advance at such a quick pace was the fact that the discoveries made as a result of exploration unquestionably benefited society, whether it be the many cutting-edge technologies and consumer products brought from China, the spices of India and Indonesia, or the crops from the Americas.[1]

The significance of that final one cannot be overstated. Crops like the potato (which acquired the nickname "ready-made bread" because it was quick to prepare) combined with maize, squashes, tomatoes, and other yams allowed European farms to sustain more humans. Potatoes are one of the most valuable sources of human and animal food during wartime (they are relatively cheap and easy to farm, easy to distribute and can last for up to several months in a cool pantry). This was also excellent for Asia, where such crops were introduced in the 17th century.

In addition, the Europeans reinforced their economy with the massive quantities of cotton, tobacco, and sugar that they "acquired" from the Americas. Slaves were the primary means by which these goods were "acquired." Slavery during European colonialism in the late second millennium is a highly emotional topic, but it is important to remember that such atrocities have occurred throughout humankind's civilisational history.[6]

The Evil Side of History

To set the scene, we must head back to around 10,000 BCE. In many parts of the world, humans have abandoned their nomadic hunter-gatherer lifestyle in favour of a more settled existence based on agriculture and animal husbandry. As the population grows and food supplies increase, those at the top tend to consolidate their power and establish an elite, leading to more inequality in society.

The appearance of writing in Mesopotamia gives the oldest definite evidence of slavery, in which slaves were either captured warriors whose families could purchase their release or indebted people who sold their freedom to cover their debts. Although wealthy and powerful households may own one or two slaves, a palace may own hundreds. The males are sent to work in the fields or on buildings, while the women are usually found in the roles of maids, concubines, or in craft industries like weaving.

However, there are several more historically ancient traces of slavery, such as in 1750 BCE, when the ruler of Babylon, Hammurabi, wrote a legal code that stipulates, among other things, how slaves must be treated and the penalties for violating their rights. Or both the Old Testament and ancient Egyptian literature discuss slavery in various forms and the treatment of slaves.

One may argue that ancient Greece was the world's first true "slave society" as the island of Chios in Ancient

Greece had a large workforce due to its emphasis on winemaking, and Greek traffickers abducted slaves from Thrace, Scythia, and Asia Minor to sell them. Some of them are used in Chios's vineyards, while the others are mainly exported from the island for sale in the Aegean (this lucrative industry is the city's backbone). Slavery eventually spread to other Greek towns, and by 400 BCE, about half of Athens' population was under slave labour!

Rome, much later, mimicked this concept and acquired slaves through military conquests and commercial incursions on so-called "barbaric border regions" (and as a further step, they were traded in slave markets). Scholars estimate that 10 to 20 % of the Roman empire's population consisted of slaves, equating to between five and ten million slaves out of an estimated 50 million Romans in the first century CE! This number of slaves would have been distributed unequally over the empire, with a more significant concentration in metropolitan regions and Italy. Slaves of different sexes and ages are valued differently, although, in the early Common Era, one slave was equivalent to around 1.5 years of salary. Possession of one was indicative of affluence and success, whereby the very wealthy Romans often owned vast domains that housed thousands of slaves.

As we previously saw, the Muslim conquests started in the 7th century, when enslaved conquest victims assisted the Arab armies and contributed significantly to their achievements and conquests. Around 700 CE, after an additional 60 years of conflict, the western shores of North Africa and the Iberian Peninsula (also called Iberia) were conquered by this support force.

About 900 CE, the Vikings of Northern Europe took control of the oceans, whereby their captives were either kept by their conquerors or sold in marketplaces (the smartest and brightest among the monastically educated fetch a higher price in Venice and Constantinople). Around the end of the first millennium CE, the Varegues, a branch of the Vikings, are in charge of a major trade route that runs from the Black Sea to Constantinople via Eastern Europe and the Dnieper River. Captured Slavic slaves are among the goods sold at this market.[6]

Additionally, following the conquest of the great Constantinople in 1453, the Ottoman Empire came to dominate the linking commercial routes between the East and the West, forcing the West to seek alternative ways to the East. One of the contributing factors to their military superiority—including the successful siege and ultimate conquest of Constantinople—was that they recruited young European Christian slaves or Janissaries, converted them to Islam, and then trained them to serve the Sultan.

In the southern Abbasid Caliphate, far beyond the mighty Constantinople, the historic city of Baghdad is where the majority of captives travelling trade routes and conquered regions wind up as slaves during this period. Following their conquest of North Africa in the late 13th century, they discovered the Sahara crossing Berber trade routes that led to the gold and slave-rich Mali empire (most captives and slaves were from animist peoples further south). Slaves are traded to Tuareg caravans, which travel through the desert for months before arriving in Cairo, Egypt. Those who make it to the end are then sold as slaves in the local marketplaces, typically becoming maids or concubines. The Trans-Saharan slave trade thrived as demand rose alongside the social status gained by slave ownership. It is believed that over a period of a thousand years, between 3.5 to 7 million slaves passed these routes.[6]

While the Trans-Saharan slave trade flourished in North Africa, the Italian republics had acquired control of the most important commercial routes in the Mediterranean by the end of the 13th century. Genoa and Venice each desired a piece of the lucrative slave trade, so they set up shop in Crimea, from where slaves were transported to the Mediterranean. It was not until the early 15th century that Portugal, which was located far to the west of Europe, began exploring the coast of Africa in search of new trade routes using the newly developed, more manoeuvrable "caravel." A group of gold mines owned by the Akan tribe was discovered by Portuguese explorers, who named the uninhabited island Sao Tomé. They arrived in the Kingdom of Kongo in 1483 and immediately established diplomatic and business ties. In little than a century, Lisbon became the wealthiest city in Europe through the trade of slaves, gold, sugar, and spices. Slaves were exchanged for European commodities like tableware and weaponry that Portugal supplied.

In 1595 in Sao Tomé, the slaves, who were now in the majority, violently revolted. Portugal regains power but prioritises its holdings in the Americas, where it has already put 300,000 slaves to work cultivating sugar cane. Following the Portuguese model, countries in Western Europe with access to the Atlantic Ocean gradually globalised the slave trade. In the Netherlands, England, and France, companies are founded to take over the new trade. Their ships were stocked

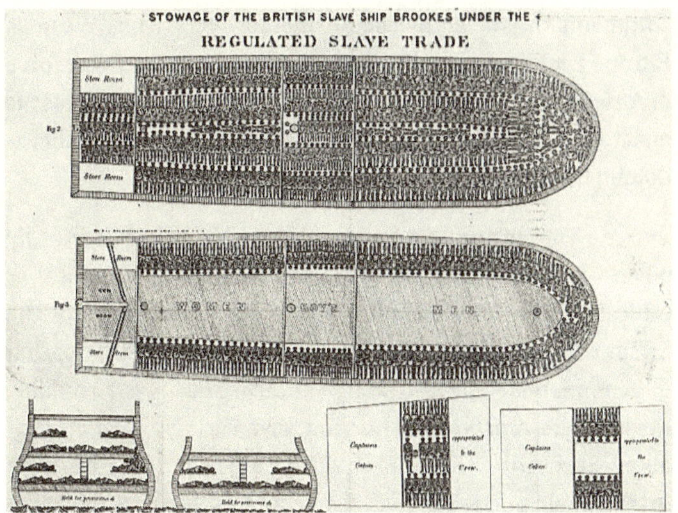

6) A plan of the British slave ship *Brookes*, showing how 454 slaves were accommodated on board after the Slave Trade Act 1788 CE. This same ship had reportedly carried as many as 609 slaves and was 267 tons burden, making 2.3 slaves per ton.

with European goods, including textiles, metallurgy, and guns. The companies set sail for the coast of Africa in search of slave markets. Included in the latter group are those who have been kidnapped or made war prisoners. Slaves, by the hundreds, are packed into the holds of ships for the months-long journey to the Americas.

The majority of these slaves wind up in Brazil and the Caribbean, where they may work in the growing agriculture industries of sugar cane, coffee, cotton, and tobacco. There, they are traded by rich European landowners, known as planters, for local goods and precious metals. Following the trade, the ships return to Europe to sell their cargo of American products. Profits are high, and the number of shipments organised by the businesses is growing. Western Europe rapidly gets richer at the expense of the labour force of Africa and the cultivable land of America. European nations enacted stringent regulations on this economic sector to defend their welfare and position.

The slave trade was largely unregulated by governments. It was a business enterprise with no moral or ethical foundations, planned and financed entirely by the forces of supply and demand in the free market. Participation in commercial slave-trading enterprises was available through the Amsterdam, London, and Paris stock markets.[0] Many Europeans from the middle class who were eager to extend their portfolios purchased these shares. With such a substantial infusion of money, companies used this capital to invest in ships, pay for the salaries of sailors and troops, buy slaves in Africa, and bring them to the New World. Slaves were sold to planters for money that was then used to buy goods from the plantations, including sugar, cocoa, coffee, tobacco, cotton, and rum. Then, they return to Europe, where they acquire a fortune in the sugar and cotton trade before setting ships for Africa yet again. Shareholders were delighted with the results: Slave trade investments returned over 6% annually during the 18th century, making them highly attractive by today's standards.[0]

As a result of the sugar industry and the rising demand for handmade goods, many new jobs were created in Europe, and many more people were able to afford consumer goods from the United States as a result of the unexpected influx of supply. Slaves in Africa became increasingly valuable as their demand rose, sparking conflicts over who will control the lucrative trade. As the kidnappings spread deeper into the region, entire communities are occasionally caught up and sold off to European ships.

Wars break out as European superpowers fight for control of the world's arable land. During the Seven Years' War, which began in 1756 and ended with the Treaty of Paris in 1763, France lost its North American possessions to Britain and Spain (but it managed to hang on to the vast majority of its sugar islands). But as Britain moved to tighten its imperial grip on the American colonies, rebels on the continent sprung up, setting the stage for what would become known as the American Revolutionary War, also known as the War of Independence. On July 4, 1776, the Continental Congress issued the Declaration of Independence, which officially declared the independence of the 13 American colonies from Great Britain.

After a disastrous expedition in 1781 results in the loss of numerous slaves who are then cast overboard, the enterprise responsible for the expedition goes to its insurer for compensation but is turned down, sparking legal proceedings in London. The public's attention and reaction to the case helped bolster abolitionist efforts. When the Society for Effecting the Abolition of the Slave Trade was established some seven years after the trial, it managed to get the first rule passed that limited the number of slaves on each ship (despite its groundbreaking success, it still had very modest influence). Similarly, the first abolitionist groups emerge in France not long afterwards, most notably with the formation of the Society of the Friends of the Blacks. However, its influence is very little in comparison to the potent slave industry.

It is estimated that as the 18th century drew to a close, about 7,700,000 Africans had been sent to the Americas as slaves. Specifically, the United Kingdom and France controlled the majority of this industry. The French colony of Saint-Domingue in the Caribbean was the richest due to its position as the world's largest exporter of sugar and coffee. However, in 1789, the French Revolution took place. "Men are born free and remain free and equal in rights," as stated in the Declaration of the Rights of Man and of the Citizen. With this statement, slaves and freemen in Saint-Domingue gained hope and inspiration and rose in rebellion. For the sake of peace and order, the French government dispatched a city commissioner. In the midst of a conflict with England and Spain, he granted the slaves freedom and got their military support (slavery was abolished formally in all French colonies the following year in Paris).

In 1807, after having sent over 2,750,000 slaves, the United Kingdom finally put an end to the slave trade, while slavery continued to be legal in its colonies. The nation is now using its influence to eradicate slavery everywhere. With Napoleon's defeat, European leaders gathered in Vienna, Austria, in 1815, and each nation made a promise to the United Kingdom that they would end the slave trade. Nevertheless, it would take decades for it to end. But in 1802, Napoleon re-established slavery after seizing power and sent an army to retake Saint-Domingue, only to be brutally crushed. Slavery was finally abolished in the United Kingdom in 1833, after which many followed. Although slavery has been banned by every nation's government, it is still practised in many parts of the world, particularly in Asia and Africa.[6] (Although every nation's government has abolished slavery, one might claim that it is being practised in some parts of the world today! There have been claims of modern slavery and child labour in cobalt mines, notably in the Democratic

Republic of the Congo, where a substantial amount of the world's cobalt is mined. The cobalt mined in the Congo will be included in every lithium-ion rechargeable battery made today).

While it did bring forth many heinous crimes, the globalisation of world zones was ultimately beneficial to our collective knowledge, which would prove to be our salvation in many ways. Consolidating global time zones did not immediately result in a revolution in how *Homo sapiens* gathered resources. The adoption of agriculture some 12,000 years ago was the last significant transition. In the second half of the second millennium CE, colonising European nations continued their agricultural traditions. These expeditions paved the way for a network of trade that ultimately resulted in the Industrial Revolution, a turning point in humanity's ability to harness more energy and generate ever-greater cultural complexity. With the advent of the Industrial Revolution, many formerly handcrafted items were mass-produced by machines, and the economy shifted from an agrarian to a manufactured one.

As we shall see in the last chapter, although fire provided *Homo sapiens* power, agriculture led to an insatiable desire for more, the creation of money gave us purpose, and the development of science made us more dangerous than ever before, the next revolution was the final true turning point for *Homo sapiens*: one that greatly expanded the complexity, potential, and dangers as a species.

12. THE INDUSTRIAL REVOLUTION

People often compare the Industrial Revolution to the Cambrian explosion—about 540 million years ago—because of the similarities in the two events in terms of the rapid increase in productivity and innovation that resulted from a relatively minor change in production methods (in the case of the Industrial Revolution, in the 1800s and 1900s and the adoption of fossil fuels). Keep in mind that evolution can continue rapidly when a new ability or characteristic enables previously inaccessible "niches" to collect energy from the environment. This biological evolution occurred during the Cambrian explosion. The speed of cultural change increased throughout the Industrial Revolution.

The term "Industrial Revolution" refers to a historical transition from an agrarian and craft economy to only one dominated by the production of machinery. Technology, socioeconomics, and culture were all crucial components of the Industrial Revolution. These technical advancements made previously impossible tasks possible, radically altering how *Homo sapiens* work, live and experience the world.

A New Ocean of Energy

Muscle power was essential for nearly all of humanity's endeavours since it was the solely accessible device for converting energy (the animal and human muscle). Muscle power from humans was used to construct boats and buildings, cattle were used to till the fields, and horses transported cargo. Plants were the ultimate source of the energy that powered these organic muscular engines. Plant growth cycles and the varying patterns of solar energy (day and night, summertime and wintertime) dominated throughout human history. When farmlands were just sown and green and daylight was sparse, society lacked energy. Kings tended to preserve the peace, law enforcers were lazy, armies had trouble moving and fighting, and the food supply was low. Sunlight and blooming grain meant it was time for the farmers to reap their harvest and stock the grain warehouses. After a successful harvest, those tasked with collecting taxes made haste to do their job, the troops stretched their muscles and readied their weapons, and the kings met in council to plot their subsequent military actions. There is no denial: Crops stored the energy of the Sun, which fed humankind.[0]

Ironically, every day for many millennia, *Homo sapiens* overlooked the single most crucial innovation in the history of energy generation that was glaring them in the face the entire time. A cook or maid sets a pot on the stove to boil water, and as soon as the water starts to boil, the lid starts to bubble, eventually flying off (energy was being generated from heat). However, lids that sprang off of pots when they were left on the stove were an inconvenience.

After the development of gunpowder in the 9th century CE in China, some progress was made in transforming heat into motion. For centuries, gunpowder was mainly utilised to produce smoke bombs since the thought of utilising it to drive projectiles seemed so counterintuitive. However, firearms did develop in time (perhaps when a weapons expert pounded gunpowder in a mortar, only to have the stone fly out with some velocity). Reliable artillery didn't appear until almost 600 years after gunpowder was discovered. Still, the concept of turning heat into motion was so counter-intuitive that it took nearly three centuries for the following mechanism to be invented that did so.

The innovative technology originated in Britain's coal mines, where it was first used when the country's expanding economy and the shrinking supply of firewood led people to switch to using coal as a fuel source instead. Frequent floods, however, prevented miners from reaching the deeper floors of the mines, where multiple coal seams were located. It remained a problem in need of a solution until approximately 1700 when a peculiar sound began echoing in British mineshafts: a steam engine. Steam engines come in a wide variety, but all of them are based on the same fundamental concept: heat-to-motion conversion. With the help of the heat generated by the combustion of a fuel like coal, water is brought to a boil, and steam is released. A cylinder or piston is pushed forward by the expanding steam. Anything attached to it will also move when it rotates.[0]

During the following decades, businessmen in Britain enhanced the performance of the steam engine, took it out of the mines, and wired it up to textile machinery like looms and gins. It allowed for the mass manufacture of affordable textiles at a lower cost, which ultimately revolutionised the textile industry. Moreover, replacing the steam engine from tunnels made a new psychological breakthrough: people began to wonder whether this technology could be used to

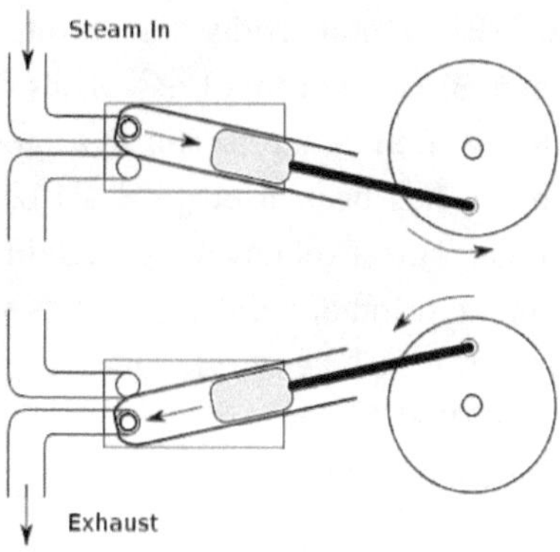

7) Operation of a simple oscillating cylinder steam engine.

move something else. While a British engineer installed a steam engine to a train of coal-filled mine carts in 1825, the first commercial railway system between Liverpool and Manchester quickly opened just five years later (on 15 September 1830). Tow barges, propelled by horsepower down the tow route, were the first kind of public transportation in use in the 17th century. Yet, these moved slowly, at 3–4 kilometres per hour, but always arrived on time. However, the railroad ushered in a new era of efficiency and productivity, and in the 19th century, steam engines achieved speeds of up to an astonishing 60 kilometres per hour.[3]

The concept that machinery and engines might be utilised to transform one form of energy into something else sparked the curiosity of humans everywhere. With the appropriate machinery, humanity could potentially use any source of energy, wherever in the world, to meet any goal. New, more effective and quicker ways of transport, cotton gins, mule spinning, power looms, and steam power all contributed to a rise in total production output, and in almost a blink of an eye, Britain became the world's first industrial powerhouse. More often than referring to a specific historical period, the phrase has been used to describe a method of economic change. This explains why certain parts of the world, like China and India, didn't start their first industrial revolutions until the 20th century, while other parts of the world, like the United States and western Europe, already started their "Second" Industrial Revolution in the late 19th century.[2]

During those years, from around 1760 to 1830, the United Kingdom was the sole epicentre of the Industrial Revolution. Noticing they had a head start, the British banned the transfer of manufacturing know-how, technology, and machinery abroad.

However, several Britons saw attractive economic opportunities elsewhere, and continental European merchants attempted to persuade British knowledge to go there (thus, the monopoly did not survive for long). Because of the work of two brave Englishmen, William and John Cockerill, who established machine shops in Liège in 1807, Belgium became the first country in continental Europe to undergo industrialisation. Iron, coal, and textiles were at the heart of the Belgian Industrial Revolution, just as they had been during its British forerunner.

Comparatively, France's rate of industrialisation lagged behind that of both the United Kingdom and Belgium. While Britain was developing its industrial supremacy, France was in the midst of its Revolution, and the unpredictability of the time deterred investors from making significant financial commitments to technological advances.

Remember King Louis XVI, who discovered that paying interest on his debts was eating up about half of the yearly budget and that France was going to be flat broke? Not long after, towards the end of the 18th century, France was on the verge of bankruptcy due to its expensive role in the American Revolution and King Louis XVI's lavish spending.

Years of bad harvests, drought, cattle disease, and increasing food costs stoked unrest among the people and the urban poor, and the royal coffers were completely broke. Many people took their desperation to the streets in acts of violence and looting as a way to vent their frustration with a government that had repeatedly refused to alleviate their economic hardships.

Beginning in 1789 and ending with Napoleon Bonaparte's rise to power in the late 1790s, the French Revolution was a cataclysmic event in world history. During this time, French citizens overthrew centuries-old political institutions such as the monarchy and the feudal system. Both King Louis XVI and his queen, Marie Antoinette, were executed by the guillotine because of public disapproval of their economic policies and the French nobility. Despite degenerating into a slaughter during the Reign of Terror, the French Revolution was essential in shaping contemporary democracies by demonstrating the power of the will of the people.

Frustration with the Directory's rule reached a boiling point on November 9, 1799, prompting Napoleon Bonaparte to execute *a coup d'état* (political group) and designate himself "first consul" of France. The Napoleonic period, in which France came to control most of continental

Europe, began with this event, marking the end of the French Revolution.

The French Revolution contributed to the spread of nationalism across Europe by radically altering the political structure of France, outlining the rights of its inhabitants, and creating a doctrine of distinctive national symbols. Other countries' nationalist sentiments were also disseminated by the Revolution.

Yet, nationalism can act like a double-edged sword. On the one side, nationalism can be exercised positively and inclusively. Independence movements all around the world could not have succeeded without nationalism. The result was the establishment of modern nations and a complete reshaping of the world. National pride, however, may easily morph into bigotry and violence. As a result of nationalism, enormous areas of the world were colonised, and many destructive wars were fought. Further, Adolf Hitler's evil tendencies and the destruction of World War II were fueled by nationalism.

European warfare was altered by mass conscription and total war, which raised an unprecedented level of manpower and mobilised the economic and industrial might of whole nations in support of the war effort. But throughout the early stages of the Industrial Revolution, it became much simpler to mass-produce weapons and, hence, equip larger armed groups. After the French Revolution, the country shifted its attention to the industrialisation of its industry, and by the year 1848, France became an industrial powerhouse. Rapid growth was seen over the Second Empire, but that didn't prevent it from falling well behind Britain. The French Revolutionary Wars, which included the War of the First

8) *The Battle of the Pyramids* by Louis-François Lejeune. This battle was part of the French Campaign in Egypt and Syria during the War of the Second Coalition.

Coalition (1792–1797) and the War of the Second Coalition (1800–1805), were sparked by lingering disagreements over disputed territories during the French Revolution. In addition to the First and Second Coalitions, a common breakdown of the Napoleonic Wars includes the Third Coalition (1803–1806), the Fourth Coalition (1806–1807), the Fifth Coalition (1809), the Sixth Coalition (1813–1814), and the Seventh Coalition (1815), along with the Peninsular War (1807–1814) and the French invasion of Russia (1812).

However, remarkably, by 1860, even with the advent of industrialisation, the scale and extent to which the British economy relied on slave labour had reached new levels. The Lancashire mills and the 465,000 textile workers who worked in them were a showcase of the Industrial Revolution, but they could not have succeeded without the efforts of three million cotton slaves in the American South. However, the Industrial Revolution reduced the demand for human labour in production by making formerly labour-intensive tasks easier to automate. In turn, this increased output meant less need for slaves and other forms of paid physical labour.

However, coming back to the industrialisation of the world, in contrast, the rest of Europe was quite slow to catch up with the industrialisation process. In comparison to its counterparts in Britain, France, and Belgium, their bourgeoisie (the middle class) lacked wealth, power, and opportunity. There was a lack of political stability in other countries, which stifled industrial development there as well. For instance, Germany did not start its industrial growth until after national unification was attained in 1870, despite the country's abundant coal and iron resources. Germany's industrial production took off so quickly once it got going, and by the end of the century, Germany was producing more steel than Britain and had taken the lead in the chemical industries worldwide. The growth of American industry in the 19th and 20th centuries likewise surpassed European endeavours of the time—meanwhile, Japan's participation in the Industrial Revolution was also notable for its success.

In the early 20th century, eastern European countries lagged behind the rest of Europe. The industrialisation that took Britain a century and a half was compressed into a few decades thanks to the five-year plans, which allowed the Soviet Union to become a great industrial force (of course, it's arguable that powers that adopted Britain's industrialisation framework later needed less time and effort to get it up and functional). In the middle of the 20th century, countries like China and India—which had traditionally been unindustrialised—began to experience an industrialisation boom of their own.

Societal and cultural standards had to adapt to the new technical and economic reality of the Industrial Revolution. The 19th century brought no improvement in the ethics of capitalism, and the Industrial Revolution that swept through Europe enriched the bankers and capital owners but condemned millions of workers to a life of abject poverty. At first, it began to look to make workers' lives even more miserable by increasing their level of poverty. They had to rely on expensive, hardly owned production equipment for their jobs and survival, and there was a lack of stability in the workplace, as workers were regularly laid off due to technological advancements and an excess in the available workforce. Long hours for little pay, poor housing conditions, and workplace abuse and exploitation were the norm before the advent of modern labour rights. But just as new problems arose, new approaches to solve them emerged. These concepts prompted inventions and rules that improved people's material conditions, allowing them to create more and faster.

Particularly in the years after World War II, fear of Communism helped curb capitalist greed. However, injustice is still all around us. Regardless of the fact that the developing economy of the 21st century is much bigger than that of 1500, many impoverished labourers come home at the end of the day with less bread than their forebears did centuries before. Just as the beginning of the Agricultural Revolution turned out to be a huge hoax, the growth of the modern economy may be an elaborate ruse. Widespread hunger and terrible poverty may worsen as the human population and the global economy both rise. Although the Agricultural Revolution circa 12,000 years ago is worthy of lament, by that time, it was too late to give up farming. In a similar manner, one could dislike capitalism, yet we absolutely need it in order to survive. Nonetheless, can the economy grow in size indefinitely?

The Technical Revolt

New technologies, such as the internal combustion engine, electricity, and the telephone, significantly increased productivity and revolutionised transportation and communication during the Second Industrial Revolution, also known as the Technological Revolution, which took place in the late 19th and early 20th centuries.

There was widespread belief in this "new" Industrial Revolution in the late 19th and 20th centuries, despite substantial overlap with the "old." As for raw materials, the modern industry began to make use of a wide variety of previously untapped natural and synthetic resources, including lighter metals, rare earths, new alloys, synthetic goods like plastics, and alternative energy sources.[2] It took over a generation for the internal combustion engine to effectively transform petroleum into a liquid kind of political power and revolutionise how people and goods travelled. For millennia, *Homo sapiens* have been using petroleum for things like lubricating wheels and making rooftops waterproof. But no one realised its potential beyond that usage until the last century.[0]

The lack of raw resources was the second issue that slowed economic expansion until *Homo sapiens* learned how to properly harness and transform energy. When humans figured out how to generate vast amounts of extremely inexpensive energy, they were able to access and utilise previously inaccessible sources of raw materials, or transport raw resources from ever more remote regions. At the same time that we were learning about previously undiscovered natural elements like silicon and aluminium, we were able to develop brand-new raw materials like plastics.

Applied science opened many opportunities. Midway through the 19th century, there existed a scientific grasp of chemistry and a fundamental understanding of thermodynamics, and by the latter quarter of the century, both of these disciplines were close to their present-day fundamental form. In developing physical chemistry, thermodynamic concepts were utilised. Understanding chemistry was crucial in the creation of the fundamental inorganic chemical and aniline dye industries. This innovation paved the way for common materials such as nylon, polyester, acrylic, spandex, olefin, synthetic fur, leather, and suede (something you'll certainly find in your everyday items).

A "second" Industrial Revolution in transportation began in 1886, when German inventor Karl Benz created the first automobile using newly available materials. Unlike traditional carriages, which use wooden wheels, his vehicle had wire wheels, as well as a four-stroke engine with a very sophisticated coil ignition and evaporative cooling in place of a radiator. Two roller chains transferred power to the back wheels. In the late summer of 1888, Benz began selling the vehicle (marketed as the Benz Patent Motorwagen), making it the first car to be sold to the general public.

About a decade later, Henry Ford created his first automobile in 1896, and he, along with other industry pioneers, worked together until 1903, when they established the Ford Motor Company. Henry Ford envisioned a mass-

9) An advertisement in 1903 of Ford's first production car, the Model A, with an under-the-floor engine selling for $850. At the time, it was (as the advertisement states) "the boss of the road" and "the newest and most advanced auto".

produced automobile that would be cheap to the ordinary worker, and Ford and his colleagues at the firm tried to find ways to increase output in line with this goal. Scientific management, often known as Taylorism, was created by Frederick Winslow Taylor and others in the United States during the Second Industrial Revolution.

With the help of Taylor's principles, Ford Motor Company reorganised their facility, placing machine tools and special purpose equipment in a logical order to facilitate the production process. Mass production refers to the elimination of all non-essential human motions by arranging all work and tools in close proximity to one another, and where practicable, on conveyors to create the assembly line. For the first time ever, hundreds of thousands of massive, complicated products with 5,000 individual pieces may be manufactured in a single year. The Model T's price dropped from \$780 in 1910 to \$360 in 1916 because of cost-saving mass production techniques. In 1924, Ford manufactured two million T-Fords, selling them between \$260 and \$850 each (nominal dollars, equivalating to about \$5,000-10,000 in today's money).

The groundbreaking T-Ford soon gained notoriety thanks to the attention it received in the press. But the easiest method to spread the message or flaunt your new high-tech purchase was through the telegraph. In May of 1837, Sir William Fothergill Cooke and Charles Wheatstone set up a telegraph line between London's Euston station and Camden Town. Throughout the century, telegraph networks rapidly expanded, with John Watkins Brett laying the first undersea telegraph cable between France and England. In 1856, London saw the need for a commercial telegraph wire across the Atlantic Ocean, and the Atlantic Telegraph Company was born to meet that need. Despite several setbacks, on July 18, 1866, the SS Great Eastern, commanded by Sir James Anderson, finally made it to its destination. The British underwater cable system was the most advanced in the world from the 1850s until 1911. However, shortly thereafter, another new communicative technology would be praised.

In the late 19th century, a young man by the name of Guglielmo Marconi, born in Italy, grew up with an interest in science and technology. The concept of wireless communication captivated him, and he was determined to discover a method to make it a reality. Marconi had an epiphany one day. He recognised that the invisible electromagnetic waves that surround us might be utilised to carry messages over vast distances. Numerous scientists and engineers believed it was impossible to transmit wireless messages over substantial distances using electromagnetic waves. They believed that these waves could only travel in a straight line and that mountains, buildings, and trees would obstruct their path. However, Marconi believed it to be achievable and set out to prove everybody wrong. In 1895, he began experimenting at his father's home in Pontecchio, and by changing the antenna's design, he was soon able to transmit signals across a distance of many kilometres. Marconi formulated a concept that became known as "Marconi's law," which describes the correlation between antenna length and the maximum distance a signal may travel.

Marconi's innovation, a radio transmitter and receiver, was simple yet effective. Without the need for connections or wires, people may communicate messages wirelessly using this device. It was a groundbreaking concept that instantly captured the attention of people worldwide. The innovation of Marconi swiftly made its way into ordinary life. Communication between ships at sea and with shore-based stations would make navigating safer and easier. News may be delivered instantaneously, linking individuals in previously unimaginable ways.

Moreover, Marconi's innovation cleared the way for the current world of communication, linking individuals all over the globe and facilitating the easy exchange of ideas and information. (Consider how revolutionary that was. Almost 2,000 years ago, a messenger might have travelled within the centre of the Roman Empire in less than a week, but it would have taken more than a month to reach the empire's borders. Yet, only because of Marconi's breakthrough, this would be practically instant!).

Marconi was celebrated as a hero and visionary. He was acknowledged as one of the greatest innovators of all time and was awarded a number of honours and medals for his groundbreaking innovation. His legacy continues to live on as we continue to use and enhance his wonderful innovation. Remember the name Guglielmo Marconi and the amazing influence he had on the world the next time you use your radio, mobile phone, or any other type of wireless communication. With his great mind and unrelenting zeal for science and technology, he genuinely altered the direction of history.

A Shocking Upheaval

In 1850, William Gladstone, the British statesman, posed a question to Michael Faraday, a physicist and pioneer in the field of electricity, why electricity was valuable, to which Faraday answered, "One day, sir, you may tax it." Faraday created the theoretical and practical groundwork for the utilisation of electrical energy. Faraday's studies of the magnetic field surrounding a conductor carrying a direct current laid the groundwork for the notion of the electromagnetic field in physics.

Even in the industrialised West, most people did not have access to electricity until far into the 20th century. As a result, it was frequently portrayed as a mysterious, quasi-magical power that could slaughter the living, raise the dead, and otherwise bend the rules of nature in popular culture of the period. The experiments of Luigi Galvani in 1771, in which he demonstrated that applying galvanism (the generation of

10) Built in 1831 CE, the Faraday disc was the first electric generator. The horseshoe-shaped magnet (*A*) created a magnetic field through the disc (*D*). When the disc was turned, this induced an electric current radially outward from the centre toward the rim. The current flowed out through the sliding spring contact *m*, through the external circuit, and back into the centre of the disc through the axle (Faraday discs are not designed for electrical energy storage, so unfortunately you cannot charge your mobile phone with them!)

electric current within biological organisms) could cause the legs of dead frogs to twitch, marked the beginning of this perspective. Shortly after Galvani's research, accounts of the "revitalisation" or resuscitation of people who had been declared clinically dead or drowned began to appear in the medical literature. When Mary Shelley wrote *Frankenstein* (1819), she was aware of these outcomes, albeit she didn't specify the mechanism used to reanimate the monster. Eventually, the concept of reviving monsters with electricity became a standard plot point in the horror genre.

Whether they were fictional or real geniuses like the mathematician and electrical engineer Charles Steinmetz, inventor and businessman Thomas Alva Edison, or inventor and electrical engineer Nikola Tesla, those who mastered electricity were often portrayed as wizards.

Alternating current (AC) electrical systems owe a great deal to the work of Nikola Tesla. AC motors and generators, which he pioneered, are utilised in everything from power generation to electric cars to home appliances. The Tesla coil, an amazing high-voltage and high-frequency transformer, is one of Tesla's most notable achievements. This device has several practical uses, including wireless communication and the generation of X-rays. Tesla contributed much to the advancement of radio technology and has numerous important patents in this area. His thoughts on the many applications of electricity and his hope for a sustainable future are equally well-known. In sum, Tesla's contributions to electrical engineering and technology have had and continue to have far-reaching effects, shaping the way *Homo sapiens* live and work in the present day.

As a result, the field of electricity was an extremely important one. Back when it was invented two centuries ago, electricity did not play a part in the economy. At most, it was utilised for obscure scientific studies and low-cost magic performances. It was transformed into the one and only genie in the lamp by a slew of innovations. We give it a flick of our fingers, and it prints this book and stitched dresses, transports humans in public transport, preserves our food, prepares our meals, turns on our lights at night and provides us with endless comedy shows and YouTube videos to watch. Even fewer of us are able to conceptualise what our lives would be like if we did not have access to electricity, along with all of its benefits.

The transformation of energy has been the driving force behind most of the progress made throughout the Industrial Revolution. It has repeatedly shown that the quantity of energy that we have access to is beyond bounds, which is a significant revelation. Or, to put it another way, the only limit is that which is imposed by human ignorance. We seem to find a new source of energy once every few decades, which means that the overall amount of energy that we have access to is always increasing.

The Industrial Revolution made possible the unprecedented availability of cheap energy and easily accessible raw materials, leading to a huge boost in human productivity. One of the first sectors hit by the eruption was agriculture. Common pictures associated with the Industrial Revolution include smoke-belching chimneys dotting the horizon and sweaty, oppressed coal miners toiling away underground. However, the Industrial Revolution could not have occurred without the preceding Second Agricultural Revolution. Industrial production techniques have been the norm in agriculture for the past two centuries. Things that had to be done by hand or weren't done at all because no one had the means to accomplish them were now being done by machines like tractors. The use of hormones and medicines, in addition to synthetic fertilisers and industrial pesticides, has greatly increased the output of farmland and livestock. The development of refrigeration, container shipping, and aeroplanes made it possible to store fruit for extended periods of time and export it to the other side of the world at a low cost and in a short amount of time.[0]

Approximately 50 billion (agricultural) animals are slaughtered every year, and tens of billions more are kept as part of a mechanised manufacturing line.[2] A significant rise in agricultural output and food storage has resulted from these industrial livestock practices. Industrial animal husbandry, together with mechanised crop production, underpins the whole contemporary economic and social system. Only 2% of Americans make a living from farming today, yet their output is sufficient to feed their entire nation and even provides significant surpluses for global export.

The Industrial Revolution provided new methods of energy conversion and product production, greatly reducing humankind's reliance on natural resources. Humans have cleared land for farming, drained marshes, blocked waterways, swamped fields, constructed tens of thousands of kilometres of rail tracks, and developed sprawling urban centres with towering skyscrapers. Ecosystems were altered, and species went extinct, as the planet was reshaped to suit the demands of *Homo sapiens*.

Degradation of the environment is distinct from a lack of resources. It was established in the prior chapter that humanity's access to resources is growing and is expected to continue doing so. That's why Harari claims that it is unlikely that predictions of impending resource depletion are accurate. However, genuine concerns about environmental deterioration are well-founded. *Homo sapiens* may one day dominate an abundance of new resources and energy sources while damaging the planet's remaining natural ecosystem and causing the demise of numerous other species. In fact, ecological upheaval may pose a threat to the very existence of *Homo sapiens*. It's possible that humans may generate more and more harmful side effects if they continue to utilise their power to fight against natural forces and bend the environment to their will. It seems likely that only extremely dramatic modifications of the environment that could lead to additional severe disorders would've been capable of reining these in.

This method has been dubbed "the devastation of nature" by many. To counter this, many scholars argue that this is not annihilation but rather transformation. The natural world is untouchable. As we witnessed at the end of the book's first part, an asteroid 65 million years ago wiped out the dinosaurs, but fortunately also paved the way for mammals. Thus, humanity is only here today for the extinction of countless species and may very well destroy itself in the process. Nonetheless, reports of our own demise are well off the mark.

Population growth has been unprecedented since the Industrial Revolution. Just 18,500 breeding human ancestors were alive 1.2 million years ago. Yet, a few million *Homo sapiens* called the Earth their home at the beginning of the Agricultural Revolution (10,000 BCE). There were about 190 million of us in the world in 1 CE, and it was only at the beginning of the 19th century (1804 CE) that the world contained a billion *Homo sapiens*.

From that moment on, following the Industrial Revolution, the population size exploded exponentially: 123 years later, in the year 1928 CE, the population size had doubled to two billion; In the year 1975, just 47 years after the two billion mark, it had doubled again to four billion; And today, not even half a decade after the four billion mark, more than eight billion unique, creative and yet Stone Age-wired *Homo sapiens* are roaming the Earth! (And if *Homo sapiens* don't destroy themselves this century, which you know is

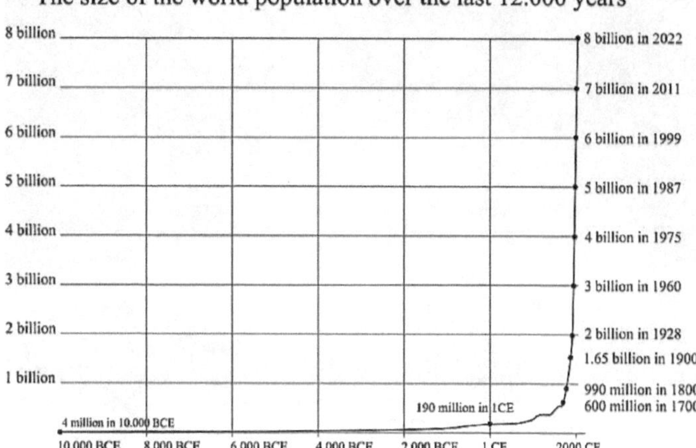

11) Around 108 billion *Homo sapiens* have ever lived on our planet. This means that today's population size makes up about 7.5% of the total number of *Homo sapiens* ever born!

hard to predict given our intrinsic nature, the world population is projected to reach 9.8 billion in 2050 and 11.2 billion in 2100).

The Modern Revolution

But despite every significant event that has been said about the Scientific and Industrial Revolution, the single most important and defining event of the previous 500 years occurred on July 16, 1945. On that day, at precisely 5:29:45 a.m., American scientists detonated the world's first atomic bomb in Alamogordo, New Mexico, as part of the Manhattan Project, a World War II effort to create nuclear weapons. From

12) The Trinity test of the Manhattan Project was the first detonation of a nuclear weapon. J. Robert Oppenheimer, wartime chief of the Los Alamos Laboratory and commonly acknowledged as "the father of the atomic bomb," famously stated, "Now I am become Death, the destroyer of worlds," after witnessing the explosion.

then on, *Homo sapiens* had the ability to not only change the world but also annihilate it altogether.

Huge shifts have occurred in society because of the Manhattan Project. At 8.15 a.m., August 6, 1945, the first ever nuclear weapon was detonated over the Japanese city of Hiroshima by an American B-29 bomber during World War II (1939-1945). Approximately 80,000 people died as a direct result of the explosion, and many more lost their lives as a direct result of radiation illness in the days and weeks that followed. When a second atomic bomb was dropped by a B-29 on Nagasaki three days later, it caused the deaths of around 40,000 people. During a radio address on August 15, the Emperor of Japan, Hirohito, announced Japan's unconditional surrender in World War II. He cited the destructive power of "a new and most cruel bomb" as the reason for Japan's surrender.

The Second World War, also called World War II, culminated in the formation of two competing military coalitions, the Allies and the Axis powers, including the vast majority of the world's nations and all of the great powers. The escalation of geopolitical tensions and conflicts that occurred in the wake of World War I, as well as the advent of authoritarian and fascist governments in Europe and Asia, were the seeds that eventually grew into World War II.

A connection may be found, however, between the First World War and the Second World War in order to have a better understanding of both of these catastrophic wars. The assassination of Archduke Franz Ferdinand, the heir presumptive to the throne of Austria-Hungary, on June 28, 1914, was the spark that set off a chain reaction of events that ultimately led to World War I. The battle escalated rapidly as governments throughout the world sent in their armed forces and declared war on one another. Europe was where the conflict broke out first, but it quickly expanded to other regions, including Africa, the Middle East, and Asia. Machine guns, poison gas, and tanks were only some of the most sophisticated weaponry and tactics used during World War I. The first widespread employment of aeroplanes in combat also occurred during this time. Many millions of people perished, and many nations' industries and infrastructures were wiped out as a direct result of the conflict.

After more than four years of horrific fighting and the loss of millions of lives, the signing of the Armistice on November 11, 1918, which officially ended hostilities between the Allied Powers and the Central Powers, marked the official end of World War I. By signing the Treaty of Versailles in 1919, the Allies were able to publicly declare victory and impose a number of "harsh" conditions on the Central Powers, who had been defeated. A period of political and social upheaval that would last for most of the 20th century began in the wake of the war, which witnessed the fall of numerous empires and the rise of new nation-states.

The shift from the remnants of the Ottoman Empire to the emergence of the modern nation of Turkey is one noteworthy example. Despite efforts to modernise, the Ottoman Empire experienced what most historians describe as "a lengthy, gradual collapse" over its 600-year existence. The majority of historians believe that being too agrarian and not unified enough, facing a disastrous rivalry with Russia, and choosing the wrong side in World War I were the roots of collapse. After fighting on the side of Germany in World War I and suffering defeat, the empire was dismembered by treaty and ended in 1922, when the last Ottoman Sultan, Mehmed VI, was deposed and departed Constantinople (now Istanbul) on a British warship.

However, the First World War was merely the beginning of the development of contemporary state nations. Immediately following the end of the First World War, a new wave of evil and hate began to brew (unnoticed by many). After suffering defeat, Adolf Hitler claimed that Germany's military was weak because of the provisions of the "harsh" Treaty of Versailles. He said that Germany would have preferred to maintain the status quo if all other countries had been disarmed. German defence preparations were necessitated since this had not occurred. Hitler promised to nullify the Treaty of Versailles in his 1924 book, *Mein Kampf*.

According to the Treaty, Germany's army size was limited to a maximum of 100,000 troops. However, a year after secretly expanding his military, Hitler addressed a massive rearmament rally in 1935, and his antisemitism was growing more extreme against the backdrop of the violent revolution. Hitler first used the germ analogy to describe Jews and stated that eliminating disease's origins was essential and believed that a "pure" German nation would be the strongest.[4]

In 1919, the German Workers' Party (Deutsche Arbeiterpartei, also named DAP), the precursor of the NSDAP (Nationalsozialistische Deutsche Arbeiterpartei, also known as the Nazi Party), was established by a Munich locksmith named Anton Drexler. Drexler was inspired by the extremist German nationalist who fought against the communist upheavals in Germany after World War I. The NSDAP was officially refounded on February 26, 1925, with Hitler as its unquestioned leader after Hitler successfully persuaded Bavarian authorities to revoke the ban on the party.[4] Until his death in 1945, Hitler's ideals, which he first began developing in the 1920s, remained largely unchanged.

But in 1933, he was given the authority to really put them into action, and shortly after Hitler's rise to power, fueled by his radical ideology, Germany invaded Poland on September 1, 1939, sparking World War II. As a result, Britain and France declared war on Germany, and the fight quickly expanded to other regions of the world. After Japan's attack on Pearl Harbor in 1941, America decided to join the war effort. Despite a modest beginning in 1939, the Manhattan Project developed to employ more than 130,000 people at the cost of about $2 billion (around $43 billion in today's money) as a direct result of the war and involvement in World War II.

13) The city centre of the Dutch city of Rotterdam after the Nazi bombing, an event known as the "Rotterdam Blitz." The heavily damaged (now restored) *St. Lawrence church* **stands out as the only remaining building that is reminiscent of Rotterdam's medieval architecture. The photo was taken after the removal of all debris.**

Roughly 70 million soldiers fought on behalf of Allied or Axis countries (the German Army had about 13.6 million soldiers), with 50–55 million civilians and about 21–25 million military killed as a result of the Second World War (including the deaths in captivity of about five million prisoners of war). While the Second World War was marked by the employment of cutting-edge weaponry and tactics, including nuclear weapons, it was also distinguished by a number of war crimes and acts of genocide, the most infamous of which was the Holocaust, in which millions of Jews and other minorities were ruthlessly exterminated by the Nazi dictatorship. As a result of the atomic bombings of Hiroshima and Nagasaki, Japan surrendered to the Allies on September 2, 1945, bringing an end to World War II. Following World War II, the United States and the Soviet Union emerged as global superpowers, leading to the outbreak of the Cold War and the decolonisation of other nations.

From the end of the war in 1945 until the fall of the Soviet Union in 1991, there was a period of political and military tension known as the Cold War between the Western powers headed by the United States and the Eastern powers led by the Soviet Union. Both sides of the Cold War fought proxy wars, in which they backed opposing factions in conflicts throughout the world as part of their battle for global power. During the course of the Cold War, both sides engaged in a nuclear arms race in an effort to prevent the other from striking.

In 1986, at the zenith of the Cold War, there were 70,300 operational nuclear weapons. However, the number of globally functioning nuclear weapons has decreased over time. Currently, nine nations possess approximately 12,700 warheads: Russia, the United States, China, France, the United Kingdom, Pakistan, India, Israel, and North Korea. The United States and Russia account for roughly 90% of the world's nuclear arsenal. Despite this, many academics believe that a global nuclear war with the current arsenal is considerably more deadly than at the Cold War peak. This horrific observation is owing to the fact that the explosive yield (the amount of energy generated when a nuclear weapon detonates) has increased tenfold since the "Fat Man" and "Little Boy" atomic bombs were unleashed on Nagasaki and Hiroshima.

Major events during the Cold War included the Berlin Blockade, the Cuban Missile Crisis, and the Vietnam War. Domestic and international politics were also profoundly affected by the conflict, as both sides worked to export their

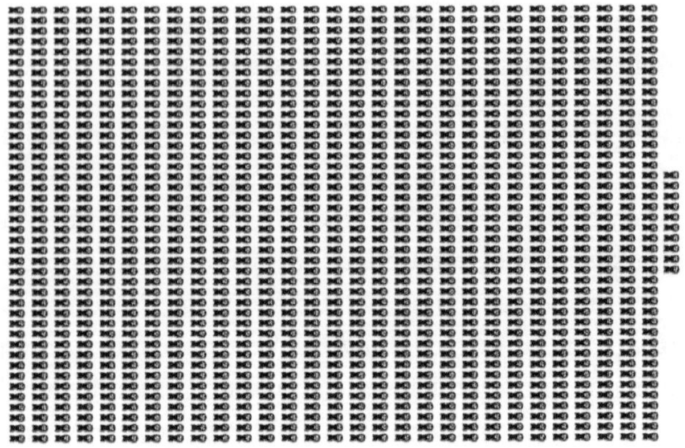

14) A concise yet clear depiction of the amount that makes up 1% of today's worldwide stockpile of nuclear weapons.

own ideologies and power. Michail Gorbatsjov, the eighth and final leader of the Soviet Union, decided to allow elections with a multi-party system and establish a presidency for the Soviet Union. This initiated a slow process of democratisation that eventually destabilised Communist control and contributed to the fall of the Soviet Union. The fall of the Soviet Union and the demise of communism in Eastern Europe marked the end of the Cold War. As the risk of escalating nuclear war decreased and the world became more interdependent, a new age of cooperation and globalisation began.

There were many significant shifts in modern civilisation caused by the Industrial Revolution. Adjusting to the pace of industrialisation is only one example. Rapid urbanisation, the decline of the peasants, the ascent of the industrial workforce, the empowerment of the average citizen, democratisation, youth culture, and the collapse of patriarchy are a few more significant examples. The disintegration of the family and the local community, and their replacement by the state and the market, is by far the most profound social shift that has ever befallen humanity, and it has occurred somewhat more recently than the previous revolutions.

Since their earliest days, hundreds of thousands of years ago, *Homo sapiens* have always clustered in connected, family-oriented communities, so far as we can tell. This truth was not affected by the Information Age or the Industrial Revolution. Tribes, towns, cities, and empires were all constructed based on reinforced communities and families. However, the Industrial Revolution was able to reduce these components to their fundamental constituents in a little over 200 years. The majority of communities' and families' traditional duties were moved to the market and the state.[0]

Bear in mind, however, the significance of World Wars I and II and the Cold War to the development of technology. Benz began selling the first mass-produced automobile in the late summer of 1888. One generation later, during World War I, the first widespread use of aeroplanes in combat and the creation of tanks occurred. Yet, in the following generation, more advanced weaponry and technology were employed in the Second World War, including more potent aircraft, ships, tanks, submarines, and the atomic bomb. In as little as a decade later, on October 4, 1957, the Soviet Union successfully launched Sputnik I, the world's first artificial satellite, kicking off the so-called "Space Race." Approximately a decade later, on July 20, 1969, millions of people around the world witnessed two brave American astronauts achieve the most remarkable milestone

15) *Earthrise*, **taken on December 24, 1968, by Apollo 8 astronaut William Anders (Apollo 8 was the first crewed spacecraft to leave low Earth orbit and the first human spaceflight to reach the Moon. The crew orbited the Moon ten times without landing, and then departed safely back to Earth). Everything we've explored so far— the story of the origin and evolution of life and humankind—took place on that tiny marble!**

in human history. Neil Armstrong and Edwin "Buzz" Aldrin became the first *Homo sapiens* to walk on the Moon, with Neil Armstrong famously stating, "One small step for man, one giant leap for mankind." as he stepped foot on the Moon.

This small step was in fact a great leap for mankind, since it revolutionised our understanding of our origins and place in the cosmos. For instance, American biologist Robert Whittaker's five-kingdom system (Monera, Protista, Fungi, Plant, and Animal Kingdom) became the new standard in science in the same year that Neil Armstrong left the first footprints on the Moon. Another example is in the 1950s, a couple of decades before the classification of nature's kingdoms, it was believed that life was less than 600 million years old. In the 1970s, scientists had a hunch that it was probably about 2.5 billion years old. However, it was only several decades ago that 3.85 billion years old became the accepted date. Or what about the fascinating stories of *Australopithecus afarensis* Lucy? Paleoanthropologist Donald Johanson from the Cleveland Museum of Natural History excavated her half a century ago, in 1974 in Africa, at a site in the Awash Valley in the Afar Triangle in Ethiopia. If we move forward a decade, it was only in the 1990s that the term "last universal common ancestor", also known as the "LUCA", was proposed. Or roughly about the same time, Grypania, the oldest known eukaryotes, were discovered in Michigan iron deposits in 1992. Or else, what about the understanding of our own species, *Homo sapiens*? In the 1990s, scholars claimed that anatomically modern humans did not exist until 50,000 years ago. Then, only several years ago, in 2017 in Morocco, after about a half-a-century debate, researchers assigned a 300,000-year-old specimen to *Homo sapiens*.

In addition to an immense influx of new scientific information, perhaps even more significant around this time was when humans made great strides toward elusive production: the Third Industrial Revolution. In the 1970s, right at the heart of the Cold War, partial automation with memory-programmable controls and computers ushered in the Third Industrial Revolution (also known as The Digital Revolution). As the Cold War exerted greater financial strain, technical progress accelerated dramatically. With the advent of those innovations, we can now fully automate a manufacturing cycle with no need for human intervention. Digital technology, such as computers, the Internet, and mobile devices, has been increasingly pervasive and integrated throughout many sectors of society and industry during this time period. How we live, work, and communicate has been drastically altered as a result of the Third Industrial Revolution.

The Internet, artificial intelligence (AI) and other forms of digital technology have climbed into the ranks of other significant human achievements like the mastering of fire and the development of the wheel. In 1989, Tim Berners-Lee, a British computer scientist, created the World Wide Web. Less than 1% of the world's technologically stored information was digital in the late 1980s, but that number grew to 99% by 2014. More than 60% of the world's population now possesses a cell phone and uses the Internet, but in 1990, just 0.25% of the world's population had a cellphone, and 0.05% were Internet users. The world has recently experienced dramatic growth in the overall amount of data generated, copied, collected, and consumed: There was a thirtyfold increase in data growth between 2010 and 2020 (from 2 zettabytes to 64.2 zettabytes. And it is predicted that by 2025, the world's data generation will have increased to 180 zettabytes). Asides from generated data, from the 1960s until now, computing capacity has roughly multiplied by a trillion, yet the philosophy and intelligence of *Homo sapiens* have remained mostly unchanged.

Pax Moderna

At the very beginning of the third millennium CE, mankind wakes up from a nightmare, stretches its arms and legs and rubs its eyes as it comes to a sudden dazzling realisation: Throughout the previous few decades, humanity has been able to tame famine, plague, and warfare.[5] Due to the fact that neither of us was alive a millennium ago, most people rarely consider this. *Homo sapiens* has forgotten how vicious and cruel the world once was. Obviously, these issues have not been entirely resolved, but they have been turned from inexplicable and uncontrolled natural forces into manageable obstacles. It was neither simple nor painless to build this modern peace, also known as Pax Moderna, by eliminating mankind's biggest foes.

Ever since the dawn of civilisation, as shown throughout history, humanity's earliest (and worst) enemy was starvation.[5] Up to recent times, the majority of humanity lived on the brink of the biological poverty threshold, below which individuals perish to starvation. A small error or bad luck may simply result in the death of a whole family or entire community. If you open a history book, you will be sure to find horrifying tales of starving populations driven by famine.

The Great Chinese Famine (1959-1961) was caused by a mix of circumstances, including drought, floods, and the Chinese government's economic policies. It is believed that between 15 and 45 million people perished as a result of the famine. A combination of causes, including a storm that damaged crops and British policies that diverted food to the war effort, led to the Bengal Famine of 1943. It is believed that between 1.5 to 4 million people perished as a result of the famine. The Irish Potato Famine (1845–1852) was brought on by crop-destroying potato blight and British policies that aggravated the crisis. The famine is believed to have caused approximately half a million deaths and led to a massive migration wave from Ireland. Stalin's agricultural policies, notably forced collectivisation and food requisitions, caused the Soviet Famine (1932-1933). It is believed that between the years 1931–1934, about seven million people died of starvation. Drought, civil conflict, and government actions that worsened the situation contributed to the Ethiopian Famine (1983-1985). It is believed that between 400,000 and 1,2 million people perished as a result of the famine. These examples are merely the tip of the iceberg; an entire book might be filled with their names and tales.

You are probably familiar with how it feels to skip meals. But have you ever felt how it is if you haven't eaten for days and have absolutely no idea where the next meal will come from? Likely and hopefully, you—like the vast majority of people reading this—have never felt this intense pain. Unfortunately, our forefathers knew it all too well.

Throughout the past century, technical, economical, and political advancements have strengthened the safety net separating humanity from the biological poverty threshold. Major famines still occur in some regions occasionally, although they are extremely rare and nearly invariably the result of human politics rather than natural disasters. Even if a person loses their job and all of their assets, they are unlikely to die of starvation in the majority of the world. Nevertheless, in most nations today, obesity has become a far worse problem than hunger. For the first time in recorded history, more humans die from overeating and obesity than from undereating, and more people die from old age than from diseases.[5]

The second enemy tamed by humanity was plagues and contagious diseases.[5] As we've already seen, the most infamous of these epidemics, the so-called Black Death, began in the 1330s in east or central Asia when the flea-dwelling bacteria *Yersinia pestis* started infecting humans bitten by fleas. On the backs of an army of rats and fleas, the plague swiftly spread over Asia, Europe, and North Africa, reaching the Atlantic Ocean in less than 20 years. About 75 to 200 million people perished, which was more than a quarter of the overall population of Eurasia (Using these numbers as a reference to modern-day Eurasia, about 1.35 billion humans would have been killed!).

The government was absolutely impotent in the face of the disaster. Other than organising public prayer sessions and parades, they had no clue how to stop the spreading of the disease, much less how to treat it. During the Black Death, doctors attempted to combat the plague by donning beaked masks, leather gloves, and heavy coats. Yet, the doctor's outfit did safeguard them from this danger. Not the bird mask, but the coat, gloves, boots, and hat performed the majority of the work.

The Black Death was neither a unique occurrence nor the greatest disease in human history. As we have seen, when the first Europeans arrived in the Americas, Australia and the Pacific Islands, the natives were plagued by catastrophic diseases. Unbeknownst to the explorers and settlers, they introduced new contagious diseases to which the locals lacked resistance. As a result, up to 90 % of the indigenous citizens perished! While the new Spanish conquerors were busy enriching themselves and abusing the locals, successive waves of influenza, measles, and other contagious illnesses swept through Mexico, dropping its population from roughly 20 million to fewer than two million in nearly half a century.

This was only one of the numerous examples. The Spanish Flu pandemic of 1918-1919 was caused by the H1N1 virus and spread around the world, with the homecoming of soldiers from World War I exacerbating the situation. It is believed that between 50 million and 100 million people died from the Spanish Flu, with some estimates reaching as high as 150 million. The Third Pandemic (1855-1950) originated in China and spread to other countries, including India, Hong Kong, and the United States. According to estimates, between 12 and 15 million individuals perished during the Third Pandemic. The Antonine Plague (165–180 CE) was an ancient epidemic brought about by the bacterium *Yersinia pestis* (some scientists argue it was smallpox). It is estimated that between 5-10 million people, including the Roman Emperor Lucius Verus, perished as a result of the Antonine Plague. And the Cholera Pandemics (1817-1923) were caused by the bacterium *Vibrio cholerae* and spread via contaminated food and water. The pandemics originated in India before spreading to the rest of the world. An estimated four million people perished as a result of cholera pandemics. Similar to the total number of famines, this list can go on forever.

Yet, the human species has become more vulnerable to epidemics during the last century as a result of population growth and improved transportation. As the world population has significantly increased and more *Homo sapiens* have moved to urban regions, cities have gotten more populous. This generates situations in which infectious illnesses can spread more easily and rapidly, particularly in slums and tenements, where humans live in close proximity. Modern metropolises such as London, New York or Tokyo provide viruses with far more fertile hunting grounds than medieval Constantinople or ancient Rome. As travel and transportation have grown more convenient, both people and goods are able to move more swiftly throughout the globe,

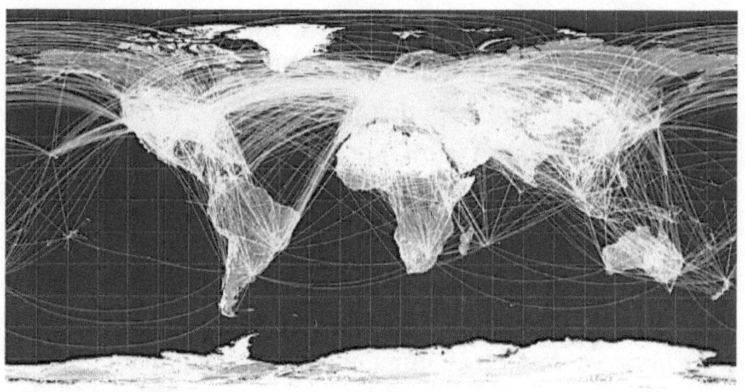

16) A map of air traffic around the world. Today, more than 100,000 flights are scheduled, facilitating the transportation of millions of *Homo sapiens* to nearly every possible location around the world. Would Columbus ever have imagined that travelling to India (or, by accident, to the Americas) would be so quick and easy?

allowing illnesses to spread rapidly across continents. A virus from one place may reach the other side of the world in less than twenty-four hours. Also, the misuse and overuse of antibiotics have contributed to the emergence of antibiotic-resistant bacteria, which are difficult to cure and easily transmissible among humans. As the majority of antibiotic-resistant superbugs kill silently between epidemics and pandemics, they have claimed more lives in the last century than epidemics and pandemics did. By 2050, the death toll from superbugs is projected to increase to ten million per year.

Every few years, a possible new epidemic emerges, such as SARS in 2002-2003, Ebola in 2014, the Zika virus outbreak (2015-2016), and SARS-CoV-2-virus (coronavirus) in 2020. Yet, owing to effective countermeasures, the number of victims has been relatively low thus far.

Many are concerned that this is really a transitory triumph and that some unidentified cousin of the Black Death or Spanish Flu is right around the corner. No one can promise that plagues will not return, but there are excellent grounds to believe that in the warfare between doctors and pathogens, doctors are winning. As we have seen in the book's first part, new diseases are mostly the product of random changes in their genetic makeup. These alterations enable diseases to transfer from animals to humans, defeat the human immune system, and develop resistance to antibiotics and other treatments. Due to human influence on the environment, such mutations likely develop and spread more rapidly today than in the past. But, in the struggle against medicine, germs are ultimately at the mercy of chance. Hence, although *Homo sapiens* will surely confront much more resistant pathogens in the following decades, treatment will likely be able to combat them more effectively than it can today.

And finally, war is the third enemy that has been tamed by humankind.[5] The majority of bloodshed throughout humanity stemmed from local quarrels between families, relatives and empires. The advent of governmental institutions is primarily accountable for decreasing violence. Ever since the beginning of the Agricultural Revolution, *Homo sapiens* living in pioneering farming communities, who were unfamiliar with state institutions, experienced a significantly high level of violence. However, as towns turned into cities, and kingdoms and empires grew in power, they repressed communities and reduced the degree of violence.

The downfall of empires, notably the British and French Empires, as well as the Soviet Union, is arguably the most prominent example. However, virtually every power we've explored in humankind's imperial history has crushed revolts with violence. And when their time came, a crumbling regime would do everything it could to rescue itself, usually resulting in a massacre. Its ultimate downfall often resulted in instability and war. However, ever since 1945, the majority of great powers have chosen peaceful withdrawal. Their fall became relatively fast, organised, and peaceful.

The fully-independent nations that succeeded these great powers were notably disinterested in conflict. Intense wars have been the hallmark of world history from the dawn of civilisation. This was how the majority of imperial powers were founded, and the majority of leaders and citizens assumed it to remain. However, wars of conquest comparable to those of Alexander the Great, Julius Ceasar or Genghis Khan don't occur anyplace in the modern world today. No UN member state has been completely eliminated or dissolved from the map by war since 1945. Of course, occasional minor conflicts worldwide still happen, but a massive violent war is no longer a viable option. According to the Law of the Jungle, violence was always a possibility in international politics, even if two sovereign entities existed in peace. Throughout the second half of the 20th century, however, this basic norm was finally violated, if not outright overturned.

In addition, a growing proportion of the human population is habituated to viewing war as inexplicable. For the first time in human history, when nations, corporations, and people analyse their immediate future, few of them consider the possibility of war. As a result of nuclear weapons transforming war between superpowers into an insane act of global suicide, the world's most powerful nations have been driven to seek out other peaceful methods of settling conflicts.

Yet, another reason is that the global economy has shifted from a material-based economy to a knowledge-based economy simultaneously.[5] Mines, fertile farmland, and oilfields were the principal sources of wealth in the past. Nowadays, knowledge is the primary contributor to wealth. Warfare can be used to seize these tangible assets, but it cannot be utilised to acquire intelligence.

Furthermore, there is no promise that this Pax Moderna component will last forever. Future technological developments may open the door for new types of conflict, similar to how nuclear weapons made the Pax Moderna viable in the first place.

Humanity has broken both the Rule of the Jungle and the Chekhov Law since 1945. Chekhov's Law, often known as Chekhov's Gun, is a theatrical theory stating that every aspect of a tale must be indispensable. This implies that every piece or detail brought into a story must serve a purpose and contribute to the broader storyline or theme. The theory was named after the Russian dramatist Anton Chekhov, who famously remarked, "If you state in the first chapter that a gun is mounted on the wall, it must fire in the second or third chapter. If it will not be shot, it has no place hanging there." In other words, everything in a story should have a purpose and add to its ultimate meaning or impact. This is applicable to all types of narrative, including literature, cinema, drama, and empirical human history. Throughout history, whenever emperors or kings acquired a new weapon, they were persuaded to use it. After 1945, however, humankind has learned to overcome this temptation. In the initial act of the Cold War, no shots were fired. As we have become accustomed to living in a world filled with undropped bombs and unfired missiles, we have become experts at breaching both the Law of the Jungle and the Chekhov Law. If these principles ever come back to haunt us, it will be our fault, not our fate.[5]

So, it seems like humanity has finally breached the imperial pattern. In general, there is finally harmony and peace. Yet, it is important to keep in mind that this position may alter in the near future (as we have seen in the timeline of world history, humans who make history can cause unexpected turns). Nevertheless, from a historical point of view, never before has peace been so pervasive that billions of *Homo sapiens* cannot even fathom famine, plague and war. Obviously, starvation, plague, and war will continue to cost millions of lives over the next few decades. The message is not that starvation, disease, and warfare have entirely vanished from the planet but that the majority of mankind need no longer be concerned about them.

Nevertheless, these calamities are no longer inescapable and beyond the comprehension and control of defenceless humankind. Instead, they have evolved into manageable obstacles. Yet, you could argue that we may infer that we are on the brink of both paradise and hell, uncomfortably wandering and balancing between the entrance to the former and the foyer of the latter. History still hasn't determined where *Homo sapiens* will eventually end, and a series of coincidental events might yet take us on any path.

But, returning to the brief history of life, at the very last, from the earliest beginnings of the birth of life in Darwin's "warm little pond", the evolution of nature's unifying forces, to ultimately the incredibly precious moment in space and time of the emergence and rise of *Homo sapiens*, to finally all the infinite pivitol revolutions and imperial crossroads, your marginal but meaningful existence with all your experiences is ultimately indelibly engraved on the tip of the roots of the Tree of Life.

PART V

The End of the Beginning

0) The James Webb Space Telescope snaps stunningly detailed image of Pillars of Creation—a formation of interstellar dust and gas located within the Eagle Nebula, about 6,500 light-years away.

AFTERWORD

13. CONCLUSION

If there is a lesson to be learned from this book, it is that all living things, including modernly articulated but yet Stone Age-wired *Homo sapiens*, have an incredible amount of good fortune just by virtue of their existence. It seems like quite an accomplishment in this Universe of ours to be able to cultivate any type of life. As human beings, of course, we have two kinds of good fortune: not only do we have the opportunity to live, but we also have the one-of-a-kind capacity to value our lives and, in a variety of different ways, to make them more enjoyable. It is a skill that we have only just started to get a grasp on. We have accomplished this feat of preeminence in an astonishingly short amount of time. Only roughly 0.0001% of the Earth's whole history has been occupied by human beings with behaviorally contemporary traits, such as the ability to communicate, create works of art, and orchestrate complicated activities. However, in order to survive even for such a short period of time, a seemingly unending line of fortunate events was necessary. We are, in fact, right at the start of everything. The challenge, of course, is to ensure that we make sure we never find the end. And to do that, *Homo sapiens* almost probably need a lot more than just a little bit of luck.

The Cambrian period of rapid growth lasted for a very long time (in the millions of years). Yet, over a period of several centuries, the Agricultural Revolution developed. We are still right in the midst of the Modern Revolution and may have just scratched the surface. Even your grandkids might not see the end of the dramatic shift in human activity and the accompanying increase in complexity during their lifetimes (Assuming, of course, that we don't act stupidly, which, you know, with *Homo sapiens*, is always a definite possibility).

Not to mention the steady increase in complexity that has occurred since the Big Bang 13.8 billion years ago. Simply said, a star is a big ball of hydrogen and helium. In contrast, an evolved brain is a complex network made up of billions of interconnected parts. The modern industrial society is a vast, whirling worldwide network of millions upon millions of minds that are more interconnected than ever. The cultural output of our civilisational consciousness added another layer of intricacy to the world.

If the earliest chapters of this book on the origin and evolution of life left you feeling small, keep in mind that because of our incredible technical advancements, mankind is currently the most complex system in the known Universe in terms of networks and building blocks. And the potential for ever-increasing complexity appears to have no limit in sight.

In the past 500 years, a staggering number of revolutions have occurred. A unified environmental and political framework has been created for the planet. The economy has flourished rapidly, and billions of *Homo sapiens* now enjoy a level of luxury that was once only attainable for legends. Undoubtedly, the Agricultural, Scientific, and Industrial Revolutions have endowed *Homo sapiens* with superpowers!

The Cognitive Revolution that transformed *Homo sapiens* from an utterly fragile and irrelevant ape into the supreme ruler of the planet did not necessitate a discernible transformation in the physiology of the human brain, not even in its size and shape. It appears to have entailed only a few minor alterations to the brain's inner core. Although opinions may vary, an additional tiny modification, such as a brain-machine interface, might be adequate to spark a Second Cognitive Revolution, produce a whole new sort of consciousness, and transition *Homo sapiens* into a completely different thing: a cybernetic organism.

The *Homo sapiens* population increased at an exponential rate during and after the Modern Revolution. While it took 200,000 years for there to be one billion *Homo sapiens*, it only took just over two centuries to achieve the eight billion mark (to put that ridiculous number into perspective, every letter in this book represents roughly 16,000 *Homo sapiens*). Right now, billions of *Homo sapiens* instantly share ideas thanks to the Internet, making the power of collaborative knowledge-building greater than ever before. Some scientists and academics have even suggested that we have passed the end of the Holocene and entered a new era called the Anthropocene because of the unparalleled influence and power that *Homo sapiens* currently have over the Earth's environment.

This era might see us taking our bit of the Universe to previously unimaginable heights of complexity, ideally to the growing advantage of all humanity and not just the fortunate few. Our great potential is a result of our group's learning efforts. Except, of course, if we reach a barrier like virtually every past power, civilisation, dynasty, empire, and kingdom. It's incredible to consider how history keeps repeating itself: In the context of life and evolution, life wants to exist, life doesn't always want to evolve, and life may go extinct at any moment, yet life keeps running. In the context of civilisational history, no power was truly universal, and from the Agricultural Revolution to the present, essentially all empires have ultimately fallen.

Our civilisational consciousness should be viewed as a precarious phenomenon comparable to a tiny candle amid a vast void. If we do not act appropriately and destroy our planet—which cannot be ruled out given the Stone Age spirit of *Homo sapiens*—before settling other worlds, this small glimmer of light will fade into the darkness of the vast and hostile Universe. Life will disappear, perhaps forever.

The Milky Way Galaxy is one of the trillions of galaxies in the observable Universe, with each galaxy holding at least hundreds of billions of stars, with most of these stars having multiple planets orbiting them. If only 0.1% of these planets host life, there must be millions of life-supporting planets in our galaxy alone. Therefore, the existence of extraterrestrial life is not only a possibility but rather a likelihood.

But will *Homo sapiens* ever achieve multiplanet status? Maybe we should discuss that briefly next time. Despite the fact that it is theoretically achievable to become a multi-planetary species, given our knowledge and resources, no one in the Universe appears to have ever accomplished this. Upon observing the night sky, we detect absolutely nothing. The Universe appears to be empty and dead. Could our hypothesised law that no reign will ever be truly universal and has a finite period also be applied in the context of every planet's dominant species' reign? Perhaps this is true, and our reign will not endure for very long. Or perhaps the future of our civilisation is bright, and *Homo sapiens* will be the first to break this law. To many, these questions may seem very lunatic, abstract or even science fiction-like. However, those who are not intrigued or spooked by these questions likely haven't given it enough thought about the principles of the staircase of life.

14. SOURCES

1. The Origin of Life

0) Bryson, B. (2004). *A Short History of Nearly Everything* (1st ed.). (PDF) (pp. 183–202) Crown.

1) Miller, Stanley L. (1953). "Production of Amino Acids Under Possible Primitive Earth Conditions." *Science.* 117 (3046): 528–9.

2) The Spark of Life." *BBC Four.* 26 August 2009. Archived from the original on 2010-11-13. TV Documentary.

3) Painter, Paul C.; Coleman, Michael M. (1997). *Fundamentals of polymer science: an introductory text.* Lancaster, Pa.: Technomic Pub. Co. p. 1. ISBN 978-1-56676-559-6

4) Denise Gellene (6 May 2013). "Christian de Duve, 95, Dies; Nobel-Winning Biochemist." *The New York Times.*

5) Oba, Y., Takano, Y., Furukawa, Y. *et al.* Identifying the wide diversity of extraterrestrial purine and pyrimidine nucleobases in carbonaceous meteorites. *Nature Communications* 13, 2008 (2022). https://doi.org/10.1038/s41467-022-29612-x

6) *Meteorites could have brought DNA precursors to Earth.* (26 April 2022). Natural History Museum, from https://www.nhm.ac.uk/discover/news/2022/april/meteorites-could-have-brought-dna-precursors-earth.html

7) Scalice, D. (2 December 2011). *NASA Astrobiology.* from https://astrobiology.nasa.gov/news/earths-early-atmosphere-an-update/

8) Miki Huynh. (5 March 2019). *NASA Astrobiology Institute* from https://astrobiology.nasa.gov/nai/articles/2019/3/5/clues-of-earths-early-rise-of-oxygen/index.html

9) Ostrander, C.M., Nielsen, S.G., Owens, J.D. *et al.* Fully oxygenated water columns over continental shelves before the Great Oxidation Event. *Nat. Geosci.* 12, 186–191 (2019). https://doi.org/10.1038/s41561-019-0309-7

10) Hodgskiss, Malcolm S. W.; Crockford, Peter W.; Peng, Yongbo; Wing, Boswell A.; Horner, Tristan J. (27 August 2019). *A productivity collapse to end Earth's Great Oxidation.* PNAS. 116 (35): 17207–17212. doi:10.1073/pnas.1900325116. PMC 6717284.

11) Suosaari, E., Reid, R., Playford, P. *et al.* (2016, February 3). New multi-scale perspectives on the stromatolites of Shark Bay, Western Australia. *Scientific Report* 6, 20557 (2016). https://doi.org/10.1038/srep20557

12) Kartik Aiyer. (2022, February 18). *The Great Oxidation Event: How Cyanobacteria Changed Life.* (n.d.). ASM.org. Retrieved October 21, 2022, from https://asm.org/Articles/2022/February/The-Great-Oxidation-Event-How-Cyanobacteria-Change

13) Lisa H. Chadwick. *Mitochondrial DNA.* (October 20, 2022). National Human Genome Institute. from https://www.genome.gov/genetics-glossary/Mitochondrial-DNA

Evolution Theory

0) Ayala, F. Jose (2022, December 3). *evolution. Encyclopedia Britannica.*
https://www.britannica.com/science/evolution-scientific-theory

1) Breitschwerdt, D., Feige, J., Schulreich, M. *et al.* The locations of recent supernovae near the Sun from modelling ^{60}Fe transport. *Nature* **532**, 73–76 (2016).
https://doi.org/10.1038/nature17424

2) Charles Darwin, *The Origin of Species*, (PDF) (chapter 2-10).

3) Zimmer, C. (2018, February 5). *The Famine Ended 70 Years Ago, but Dutch Genes Still Bear Scars.* The New York Times. https://www.nytimes.com/2018/01/31/science/dutch-famine-genes.html

The Eruption of Life

0) Bryson, B. (2004). *A Short History of Nearly Everything* (1st ed.). (PDF) (pp. 203–220) Crown.

1) Brown, J. R.; Doolittle, W. F. (1995). *Root of the Universal Tree of Life Based on Ancient Aminoacyl-tRNA Synthetase Gene Duplications.* PNAS. 92 (7): 2441–2445.
Bibcode:1995PNAS...92.2441B. doi:10.1073/pnas.92.7.2441

2) Charles Darwin, *The Origin of Species*, (PDF) (chapter 11-12).

3) Weiss, M., Sousa, F., Mrnjavac, N. *et al.* The physiology and habitat of the last universal common ancestor. *Nature Microbiology* **1**, 16116 (2016).
https://doi.org/10.1038/nmicrobiol.2016.116

4) Berggren, W. A. (2022, December 8). *Cenozoic Era. Encyclopedia Britannica.*
https://www.britannica.com/science/Cenozoic-Era

5) Malmquist, S. (n.d.). *1.7 The Evolution of Primates – Human Biology.* Pressbooks.
https://open.lib.umn.edu/humanbiology/chapter/1-7-the-evolution-of-primates/

The Dawn of Sapiens

0) Harari, Y. (2015). *Sapiens: A Brief History of Humankind* (pp. 1-70) Penguin Books

1) Crubézy, E.; Trinkaus, E. (1992). "Shanidar 1: a case of hyperostotic disease (DISH) in the middle Paleolithic." *American Journal of Physical Anthropology.* 89 (4): 411–420.
doi:10.1002/ajpa.1330890402

2) Tattersall, I. (2022, October 18). *Homo sapiens. Encyclopedia Britannica.*
https://www.britannica.com/topic/Homo-sapiens

3) Richardson, M. W. (2019, February 1). How Much Energy Does the Brain Use? from
https://www.brainfacts.org/Brain-Anatomy-and-Function/Anatomy/2019/How-Much-Energy-Does-the-Brain-Use-020119

Human Instinct

0) Winston, R. (2003). *Human Instinct* (pp. 20-192) Bantam.

1) Jerome Barkow, Leda Cosmides, John Tooby (eds), *The Adapted Mind: Evolutionary Psychology and the Generation of Culture* (pp. 249-325), Oxford University Press (1992).

2) Louise Barrett, Robin Dunbar & John Lycett, *Human Evolutionary Psychology* (pp. 45-136), Palgrave (2002).

The Tree of Knowledge

0) Harari, Y. (2015). *Sapiens: A Brief History of Humankind* (pp. 1-70) Penguin Books

1) H. B. Davis. (1907). The Raccoon: A Study in Animal Intelligence. *The American Journal of Psychology*, *18*(4), 447–489. https://doi.org/10.2307/1412576

2) Sabrina Moreyra, Mariana Lozada, Spatial memory in Vespula germanica wasps: A pilot study using a Y-maze assay, Behavioural Processes, 189, (104439), (2021). https://doi.org/10.1016/j.beproc.2021.104439

3) Gottfredson, L. S. (1997, January). Mainstream science on intelligence: An editorial with 52 signatories, history, and bibliography. *Intelligence, 24*(1), 13–23. https://doi.org/10.1016/s0160-2896(97)90011-8

4) Matthew Sims, Julian Kiverstein, Externalized memory in slime mould and the extended (non-neuronal) mind, Cognitive Systems Research, 73, (26-35), (2022). https://doi.org/10.1016/j.cogsys.2021.12.001

5) Steele, M. A., Halkin, S. L., Smallwood, P. D., McKenna, T. J., Mitsopoulos, K., & Beam, M. (2008, February). Cache protection strategies of a scatter-hoarding rodent: do tree squirrels engage in behavioural deception? *Animal Behaviour*, *75*(2), 705–714. https://doi.org/10.1016/j.anbehav.2007.07.026

6) The Key to the Success of Homo Sapiens. (2015, May 2). *History News Network.* from https://historynewsnetwork.org/article/158425

7) Timothy M. Waring, Zachary T. Wood. Long-term gene–culture coevolution and the human evolutionary transition. *Proceedings of the Royal Society B: Biological Sciences*, 2021; 288 (1952): 20210538 DOI: 10.1098/ rspb.2021.0538

8) *A Scientifically Weak and Ethically Uninspiring Vision of Human Origins.* (2022, January 28). Discovery Institute. from https://www.discovery.org/

9) Ellis, B. J., & Bjorklund, D. F. (2005). *Origins of the Social Mind: Evolutionary Psychology and Child Development* (pp 383-410). Guilford Press.

Our Last Ancestors

0) Stefan Milo. (2022, March 4). *Life And Death 3,000,000 Years Ago* [Video]. YouTube, 2022, from https://www.youtube.com/watch?v= u6kc7rEQXpI. Includes Stefan Milo. (2021, September 17). *The complex evolution of Homo sapiens - 1,000,000 to 30,000 years ago.* YouTube. from https:// www.youtube.com/watch?v=iM6LSUpanmg

1) Brain CK. New finds at the Swartkrans Australopithecine site. Nature. 1970 Mar 21;225(5238):1112-9. doi: 10.1038/2251112a0. Erratum in: Nature. 1970 Apr 25;226(5243):386. PMID: 5418243.

2) Njau, Jackson K., and Robert J. Blumenschine. "Crocodylian and Mammalian Carnivore Feeding Traces on Hominid Fossils from Flk 22 and Flk NN 3, Plio-Pleistocene, Olduvai Gorge, Tanzania." *Journal of Human Evolution*, vol. 63, no. 2, 2012, pp. 408–417., https://doi.org/10.1016/j.jhevol.2011.05.008.

3) Berger, Lee R. "Brief Communication: Predatory Bird Damage to the Taung Type-Skull Ofaustralopithecus Africanus Dart 1925." *American Journal of Physical Anthropology*, vol. 131, no. 2, 2006, pp. 166–168., https://doi.org/10.1002/ajpa.20415.

4) Wynn, J. G., et al. "Diet of Australopithecus Afarensis from the Pliocene Hadar Formation, Ethiopia." *Proceedings of the National Academy of Sciences*, vol. 110, no. 26, 2013, pp. 10495–10500., https://doi.org/10.1073/pnas.1222559110.

5) Sponheimer, M., et al. "Isotopic Evidence of Early Hominin Diets." *Proceedings of the National Academy of Sciences*, vol. 110, no. 26, 2013, pp. 10513–10518., https://doi.org/10.1073/pnas.1222579110.

6) Larson, Susan. "Did Australopiths Climb Trees?" *Science*, vol. 338, no. 6106, 2012, pp. 478–479., https://doi.org/10.1126/science.1230128.

7) Green, David J., and Zeresenay Alemseged. "Australopithecus Afarensis Scapular Ontogeny, Function, and the Role of Climbing in Human Evolution." *Science*, vol. 338, no. 6106, 2012, pp. 514–517., https://doi.org/10.1126/science.1227123.

8) Harmand, Sonia, et al. "3.3-Million-Year-Old Stone Tools from Lomekwi 3, West Turkana, Kenya." *Nature*, vol. 521, no. 7552, 2015, pp. 310–315., https://doi.org/10.1038/nature14464.

9) McPherron, Shannon P., et al. "Evidence for Stone-Tool-Assisted Consumption of Animal Tissues before 3.39 Million Years Ago at Dikika, Ethiopia." *Nature*, vol. 466, no. 7308, 2010, pp. 857–860., https://doi.org/10.1038/nature09248.

10) Pobiner, Briana L. "New Actualistic Data on the Ecology and Energetics of Hominin Scavenging Opportunities." *Journal of Human Evolution*, vol. 80, 2015, pp. 1–16., https://doi.org/10.1016/j.jhevol.2014.06.020.

11) Helm, Charles W., et al. "Interest in Geological and Palaeontological Curiosities by Southern African Non-Western Societies: A Review and Perspectives for Future Study." *Proceedings of the Geologists' Association*, vol. 130, no. 5, 2019, pp. 541–558., https://doi.org/10.1016/j.pgeola.2019.01.001.

12) Tomonaga, M., Imura, T. Efficient search for a face by chimpanzees (*Pan troglodytes*). *Sci Rep* 5, 11437 (2015). https://doi.org/10.1038/srep11437

13) Gunz, Philipp, et al. "Australopithecus Afarensis Endocasts Suggest Ape-like Brain Organization and Prolonged Brain Growth." *Science Advances*, vol. 6, no. 14, 2020, https://doi.org/10.1126/sciadv.aaz4729.

14) DeSilva, Jeremy M., et al. "Neonatal Shoulder Width Suggests a Semirotational, Oblique Birth Mechanism Inaustralopithecus Afarensis." *The Anatomical Record*, vol. 300, no. 5, 2017, pp. 890–899., https://doi.org/10.1002/ar.23573.

15) Kappelman, John, et al. "Perimortem Fractures in Lucy Suggest Mortality from Fall out of Tall Tree." *Nature*, vol. 537, no. 7621, 2016, pp. 503–507., https://doi.org/10.1038/nature19332.

16) McNutt, Ellison J., et al. "Footprint Evidence of Early Hominin Locomotor Diversity at Laetoli, Tanzania." *Nature*, vol. 600, no. 7889, 2021, pp. 468–471., https://doi.org/10.1038/s41586-021-04187-7.

17) Ayala Francisco José, and Camilo J. Cela-Conde. *Processes in Human Evolution: The Journey from Early Hominins to Neanderthals and Modern Humans*, Oxford University Press, Oxford, 2018, p. 145.

18) Hajdinjak, M., Fu, Q., Hübner, A. *et al.* Reconstructing the genetic history of late Neanderthals. *Nature* 555, 652–656 (2018). https://doi.org/10.1038/nature26151

19) Meyer, M., Arsuaga, JL., de Filippo, C. *et al.* Nuclear DNA sequences from the Middle Pleistocene Sima de los Huesos hominins. *Nature* 531, 504–507 (2016). https://doi.org/10.1038/nature17405

20) Fleagle, John G., et al. "Paleoanthropology of the Kibish Formation, Southern Ethiopia: Introduction." *Journal of Human Evolution*, vol. 55, no. 3, 2008, pp. 360–365., doi:10.1016/j.jhevol.2008.05.007

21) Vandermeersch, Bernard, and Ofer Bar-Yosef. "The Paleolithic Burials at Qafzeh Cave, Israel." Paléo, no. 30-1, 2019, pp. 256–275., doi:10.4000/paleo.4848.

22) Richter, D., Grün, R., Joannes-Boyau, R. *et al.* The age of the hominin fossils from Jebel Irhoud, Morocco, and the origins of the Middle Stone Age. *Nature*546, 293–296 (2017). https://doi.org/10.1038/nature22335

23) Pedro Soares, Farida Alshamali, Joana B. Pereira, Verónica Fernandes, Nuno M. Silva, Carla Afonso, Marta D. Costa, Eliska Musilová, Vincent Macaulay, Martin B. Richards, Viktor Černý, Luísa Pereira, The Expansion of mtDNA Haplogroup L3 within and out of Africa, *Molecular Biology and Evolution*, Volume 29, Issue 3, March 2012, Pages 915–927, https://doi.org/10.1093/molbev/msr245

24) Bergström, Anders, et al. "Insights into Human Genetic Variation and POPULATION History from 929 Diverse Genomes." *Science*, vol. 367, no. 6484, 2020, doi:10.1126/science.aay5012.

25) Beyer, Robert M., et al. "Climatic Windows for Human Migration out of Africa in the PAST 300,000 Years." *Nature Communications*, vol. 12, no. 1, 2021, doi:10.1038/s41467-021-24779-1.

26) *World before Us*, by Tom Higham, Penguin UK, 2021, pp. 11–36.

27) Groucutt, H.S., White, T.S., Scerri, E.M.L. *et al.* Multiple hominin dispersals into Southwest Asia over the past 400,000 years. *Nature* (2021). https://doi.org/10.1038/s41586-021-03863-y

The Agricultural Revolution

0) Harari, Y. (2015). *Sapiens: A Brief History of Humankind* (pp. 87-148) Penguin Books

1) Michael Balter, (2013). Farming Was So Nice, It Was Invented at Least Twice. *Science*, https://doi.org/10.1126/article.24045

2) Jared Diamond, *Guns, Germs, and Steel: The Fates of Human Societies* (New York: W. W. Norton, 1997).

3) Natalie Wolchover, (2012). Why Can't All Animals Be Domesticated? *Live Science*, https://www.livescience.com/33870-domesticated-animals-criteria.html

4) Tim Kohler, (2017). Greater post-Neolithic wealth disparities in Eurasia than in North America and Mesoamerica, *Nature*, https:// www.nature.com/articles/nature24646

5) Statista. (2022, January 14). Total U.S. wheat area planted and harvested 2001-2021, https://www.statista.com/statistics/190352 /total-us-wheat-plantings-and-harvestings-from-2000/

6) Latham, Katherine J., "Human Health and the Neolithic Revolution: an Overview of Impacts of the Agricultural Transition on Oral Health, Epidemiology, and the Human Body" (2013). Nebraska Anthropologist. 187. http://digitalcommons.unl.edu/nebanthro/187

7) Armelagos, George J., Goodman, Alan H., Jacobs, Kenneth H. 1991 Population Growth during a Period of Declining Health.

8) Population and Environment, 13(1):9-22. Eshed, Vered, Gopher, A vi, Pinhasi, Ron, Hershkovitz, Israel 2010 Paleopathology and the Origin of Agriculture in the Levant. American Journal of Physical Anthropology, 143: 121-133.

9) Goodman, Alan H. 1993 On the Interpretation of Health from Skeletal Remains. Current Anthropology, 34(3):281-288.

10) Larsen, Clark Spencer 1981 Skeletal and Dental Adaptations to the Shift to Agriculture on the Georgia Coast. Current Anthropology, 22(4):422-423. 2006 The Agricultural Revolution as Environmental Catastrophe: Implications for Health and Lifestyle in the Holocene. Quaternary International, 150(1):12-20.

11) Tayle, N., Domett, K., and Nelsen, K. 2000 *Agriculture and dental caries? The case of rice in prehistoric Southeast Asia. World Archaeology,* 32(1):68-83.

12) Ulijaszek, Stanley J., Hillman, G., Boldsen, J.L., and Henry, c.J. 1991 *Human Dietary Change. Philosophical Transactions: Biological Sciences,* 334 (1270):271-279.

13) Gat, War in Human Civilisation, 130–1; Robert S. Walker and Drew H. Bailey, 'Body Counts in Lowland South American Violence', Evolution and Human Behavior 34 (2013), 29–34.

14) Alain Bideau, Bertrand Desjardins and Hector Perez-Brignoli (eds.), Infant and Child Mortality in the Past (Oxford: Clarendon Press, 1997); Edward Anthony Wrigley et al., English Population History from Family Reconstitution, 1580–1837 (Cambridge: Cambridge University Press, 1997), 295–6, 303.

15) Pettitt, P. B., et al. "The Gravettian Burial Known as the Prince ('Il Principe'): New Evidence for His Age and Diet." *Antiquity,* vol. 77, no. 295, 2003, pp. 15–19., https://doi.org/10.1017/s0003598x00061305.

16) *Nowell, April. Growing up in the Ice Age: Fossil and Archaeological Evidence of the Lived Lives of Plio-Pleistocene Children.* Oxbow Books, 2021.

17) Agam, Aviad, and Ran Barkai. "Elephant and Mammoth Hunting during the Paleolithic: A Review of the Relevant Archaeological, Ethnographic and Ethno-Historical Records." *Quaternary,* vol. 1, no. 1, 2018, p. 3., https://doi.org/10.3390/quat1010003.

18) Stefan Milo. (2022, May 10). *Life & Death At The Height Of The Ice Age* [Video]. YouTube. https://www.youtube.com/watch?v=S2vuL3oZogc

The Unification of Humankind

0) Harari, Y. (2015). *Sapiens: A Brief History of Humankind* (pp. 181-264) Penguin Books.

1) Angus Maddison, *The World Economy,* vol. 2 (Paris: Development Centre of the Organisation of Economic Co-operation and Development, 2006), 636; 'Historical Estimates of World Population', U.S. Census Bureau.

2) Glyn Davies, *A History of Money: From Ancient Times to the Present Day* (Cardiff: University of Wales Press, 1994), 15.

3) Deshmukh, A. (2022, February 13). Mapped: *The World's Major Religions.* Visual Capitalist. https://www.visualcapitalist.com/mapped-major-religions-of-the-world/

4) Davidson, P. (2022, October 7). *Empire.* World History Encyclopedia. https://www.worldhistory.org/empire/

5) *Yuval Noah Harari: What explains the rise of humans? - TED.* (2015, July 24). [Video]. TED. https://www.ted.com/talks/yuval_noah_harari_what_explains_the_rise_of_humans/transcript

6) Van Wilgenburg E, Torres CW, Tsutsui ND. *The global expansion of a single ant supercolony.* Evol Appl. 2010 Mar;3(2):136-43. doi: 10.1111/j.1752-4571.2009.00114.x. PMID: 25567914; PMCID: PMC3352483.

7) Stephen Chrisomalis (2010). *Numerical Notation: A Comparative History.* Cambridge University Press. p. 236. ISBN 9780521878180. Retrieved 2021-02-25.

8) The A.K. Grayson, *Penguin Encyclopedia of Ancient Civilisations,* ed. Arthur Cotterell, Penguin Books Ltd. 1980. p. 92

9) Marcia and Robert Ascher, *Mathematics of the Incas–Code of the Quipu* (New York: Dover Publications, 1981).

10) Gary Urton, *Signs of the Inka Khipu* (Austin: University of Texas Press, 2003); Galen Brokaw, *A History of the Khipu* (Cambridge: Cambridge University Press, 2010).

11) *The History of Roman Numerals.* (n.d.). https://historylearning.com/a-history-of-ancient-rome/history-of-roman-numerals

Imperial History

0) Harari, Y. (2015). *Sapiens: A Brief History of Humankind* (pp. 210–232, 264–272) Penguin Books

1) Matt Baker (2020). *Timeline of World History* (PDF). UsefulCharts. ISBN: 978-0-9877293-2-3

2) Morris, Ian (October 2010). "Social Development" (PDF). *Ian Morris.* Archived from the original (PDF) on 26 July 2011. This contains supporting materials for the following book: (b) Morris, Ian (2010). *Why the West Rules—For Now.* New York: Farrar, Straus and Giroux. ISBN 978-0-374-29002-3.

3) Modelski, George (2003). *World Cities: -3000 to 2000.* Washington DC: Faros2000. ISBN 0-9676230-1-4. Figures in main tables are preferentially cited. Part of former estimates can be read at Modelski, George (12 January 2008). "The Evolutionary World Politics Homepage." *The Evolutionary World Politics Homepage.* Archived from the original on 28 December 2008.

4) J. Lawrence Angel (May 1969). "The bases of paleodemography." *American Journal of Physical Anthropology.* **30** (3): 427–437. doi:10.1002/ajpa. 1330300314. PMID 5791021.

5) Imhausen, Annette (2006). "Ancient Egyptian Mathematics: New Perspectives on Old Sources." *The Mathematical Intelligencer.* 28 (1): 19–27. doi:10.1007/bf02986998. S2CID 122060653.

6) Midant-Reynes, Beatrix (2000). *The Prehistory of Egypt: From the First Egyptians to the First Pharaohs.* Wiley. ISBN 978-0-631-21787-9.

7) van de Mieroop, M. (2007). *A History of the Ancient Near East, ca. 3000–323 BC.* Malden: Blackwell. ISBN 978-0-631-22552-2.

8) James, T.G.H. (2005). *The British Museum Concise Introduction to Ancient Egypt.* University of Michigan Press. pp 6–7. ISBN 978-0-472-03137-5.

9) McLean, J. *The Caral Civilization | World Civilization.* https://courses.lumenlearning.com/suny-hccc-worldcivilization/chapter/the-caral-civilization/

10) deMenocal, Peter B. (2001). "Cultural Responses to Climate Change During the Late Holocene." *Science.* **292** (5517): 667–673. doi:10.1126/science.1059827.

11) Bryce, Trevor (2016). *Babylonia: A Very Short Introduction.* Oxford University Press. pp. 8–10. ISBN 978-0-19-872647-0.

12) Durant, Will (1939). "The Life of Greece". *The Story of Civilization.* Vol. II. New York: Simon & Schuster. p. 21. ISBN 9781451647587.

13) Britannica, T. Editors of Encyclopaedia (2022, September 2). *Han dynasty. Encyclopedia Britannica.* https://www.britannica.com/topic/Han-dynasty

14) Brosius, Maria (2021). *A History of Ancient Persia: The Achaemenid Empire.* Wiley-Blackwell. ISBN 978-1-444-35092-0.

15) Britannica, T. Editors of Encyclopaedia (2020, August 28). *Alexander the Great's Achievements. Encyclopedia Britannica.* https://www.britannica.com/summary/Alexander-the-Greats-Achievements

16) Arzamas. (2017, May 30). *Ancient Rome in 20 minutes.* YouTube. https://www.youtube.com/watch?v=46ZXl-V4qwY

17) Britannica, T. Editors of Encyclopaedia (2022, August 20). *Middle Ages. Encyclopedia Britannica*. https://www.britannica.com/event/Middle-Ages

18) Saliba, George (1994). *A History of Arabic Astronomy: Planetary Theories During the Golden Age of Islam*. New York University Press. pp. 245–257. ISBN 0-8147-8023-7.

19) *The rise and fall of the Mongol Empire - Anne F. Broadbridge*. (2019, Augustus 29). TED-Ed. https://ed.ted.com/lessons/the-rise-and-fall-of-the-mongol-empire-anne-f-broadbridge

20) Flint, V. I. J. (2022, August 16). *Christopher Columbus | Biography, Nationality, Voyages, Ships, Route, & Facts*. Encyclopedia Britannica. https://www.britannica.com/biography/Christopher-Columbus

21) Britannica, T. Editors of Encyclopaedia (2022, August 24). *Aztec. Encyclopedia Britannica*. https://www.britannica.com/topic/Aztec

22) Britannica, T. Editors of Encyclopaedia (2022, August 26). *Inca. Encyclopedia Britannica*. https://www.britannica.com/topic/Inca

23) History.com Editors. (2021, November 10). *Francisco Pizarro traps Incan emperor Atahualpa*. HISTORY. https://www.history.com/this-day-in-history/pizarro-traps-incan-emperor-atahualpa

The Scientific Revolution

0) Harari, Y. (2015). *Sapiens: A Brief History of Humankind* (pp. 274-373) Penguin Books.

1) Green. (2014, March 11). *The Modern Revolution*. Crash Course Big History. https://cdn.kastatic.org/KA-share/BigHistory/8.0_Transcript_CC_Modernity_2014.pdf

2) Hawking, S. (2016). *A Brief History of Time* (pp. 1-16) Bantam Dell Publishing Group.

3) Osler, M. J. , Brush, . Stephen G. and Spencer, . J. Brookes (2019, November 26). *Scientific Revolution. Encyclopedia Britannica*. https://www.britannica.com/science/Scientific-Revolution

4) Heilbroner, R. L. and Boettke, . Peter J. (2022, August 18). *capitalism. Encyclopedia Britannica*. https://www.britannica.com/topic/capitalism

5) *The world of the Dutch East India Company*. (n.d.). Nationaal Archief. https://www.nationaalarchief.nl/en/explore/the-world-of-the-dutch-east-india-company

6) Geo History. (2022, October 19). *Slavery - Summary on a Map*

The Industrial Revolution

0) Harari, Y. (2015). *Sapiens: A Brief History of Humankind* (pp. 274-373) Penguin Books.

1) Kiron, M. I. (2022, April 24). *Industrial Revolution and Its Impact in Textile Industry*. Textile Learner. https://textilelearner.net/industrial-revolution-and-its-impact-in-textile-industry/

2) Britannica, T. Editors of Encyclopaedia (2022, October 27). *Industrial Revolution. Encyclopedia Britannica*. https://www.britannica.com/event/Industrial-Revolution

3) Marks, R. B. (2022). *The Origins of the Modern World: A Global and Environmental Narrative from the Fifteenth to the Twenty-First Century (World Social Change)*. (pp. 97-126) Rowman & Littlefield Publishers;

4) A+E Networks EMEA. (2022, December 14). *World War II*. HISTORY. https://www.history.com/topics/world-war-ii/world-war-ii-history

5) Harari, Y. N. (2016). *Homo Deus: A Brief History of Tomorrow* (pp. 1-22) Random House.

IMAGE CREDITS

Part I

0) Will Ratcliff and Mike Travisano, National Science Foundation.
https://www.nsf.gov/news/news_images.jsp?cntn_id=122828&org=NSF

1) Encyclopædia Britannica, Inc. https://cdn.britannica.com/93/118093-050-9AD8F7E6/histones-proteins-DNA-units-chromatin-fibre-nucleosomes.jpg

2) Woods Hole Oceanographic Institution.
https://www.whoi.edu/oceanus/feature/the-discovery-of-hydrothermal-vents/

3) Kristina D.C. Hoeppner (© CC BY-SA 2.0).
https://www.flickr.com/photos/4nitsirk/11902636365/

4) Ali Zifan (© CC BY-SA 4.0). https://en.wikipedia.org/wiki/File:Prokaryote_cell.svg

5) Mariana Ruiz Villarreal.
https://en.wikipedia.org/wiki/File:Endomembrane_system_diagram_en.svg

6) Debivort (© CC BY-SA 3.0).
https://en.wikipedia.org/wiki/File:Coquina_variation3.jpg

7) Samuel Velasco/Quanta Magazine: Ratcliff Lab, Georgia Tech.
https://www.quantamagazine.org/single-cells-evolve-large-multicellular-forms-in-just-two-years-20210922/

8) KaiserScience. https://kaiserscience.files.wordpress.com/2015/01/finches-color.jpg

9) Stig Nygaard (© CC BY-2.0).
https://www.flickr.com/photos/stignygaard/42875169812/

10) Elembis (© CC BY-SA 3.0).
https://en.wikipedia.org/wiki/File:Mutation_and_selection_diagram.svg

11) Picture from John Romanes': *Darwin and after Darwin: An exposition of the Darwinian theory and a discussion of post-Darwinian questions - I The Darwinian theory.* P. 56, Fig. 5. Chicago, 1892. biodiversitylibrary.org

12) NASA. https://photojournal.jpl.nasa.gov/catalog/PIA21424

13) Woese, C. R., Kandler, O., & Wheelis, M. L. (1990). Towards a natural system of organisms: proposal for the domains Archaea, Bacteria, and Eucarya. *Proceedings of the National Academy of Sciences of the United States of America*, 87(12), 4576–4579. (© CC BY-SA 4.0). https://doi.org/10.1073/pnas.87.12.4576

14) Daderot, Thomas Bresson, Daderot, James St. John, Peter Halasz, Daderot (© CC BY 4.0). https://en.wikipedia.org/wiki/File:Trilobita_Diversity.png

15) Kious, Jacquelyne; Tilling, Robert I.; Kiger, Martha, Russel, Jane. *This Dynamic Earth: The Story of Plate Tectonics.* (Online ed.). Reston, Virgina, USA: United States Geological Survey. ISBN 0-16-048220-8.
http://pubs.usgs.gov/gip/dynamic/historical.html

16) DiBgd (© CC BY-SA 3.0). https://en.wikipedia.org/wiki/File:Pederpes22small.jpg

17) Pixabay. https://pixabay.com/nl/illustrations/dinosaurus-tyrannosaurus-dier-t-rex-6286030/

18) Tsaiproject (© CC BY-SA 3.0).
https://en.wikipedia.org/wiki/File:Meteor_Crater_Panorama_near_Winslow,_Arizona,_2012_07_11.jpg

19) Matteo De Stefano / MUSE-Science Museum (© CC BY-SA 3.0).
https://en.wikipedia.org/wiki/File:Plesiadapis_sp._-_MUSE.JPG

20) Fred the Oyster (© CC BY-SA 4.0).
https://commons.wikimedia.org/wiki/File:Hominidae_chart.svg

Part II

0) EU (© CC-BY-SA 3.0).
https://upload.wikimedia.org/wikipedia/commons/1/1e/Lascaux_painting.jpg?20071002154413

1) Smithsonian Institution. https://humanorigins.si.edu/

2) Smithsonian Institution. https://humanorigins.si.edu/

3) Smithsonian Institution. https://humanorigins.si.edu/

4) Smithsonian Institution. https://humanorigins.si.edu/

5) Phys.org. https://phys.org/news/2015-11-nature-nurture-human-brains-evolved.html

6) Encyclopædia Britannica, Inc. https://cdn.britannica.com/73/91873-050-F952750D/stone-tools-edges-point-flakes-method-hammer.jpg

7) Dennis O'Neil. https://www.palomar.edu/anthro/hominid/australo_2.htm

8) Edublox. https://www.edubloxtutor.com/amala-kamala/

9) The Book of Threes. https://www.bookofthrees.com/triune-brain/

10) Tero, A., Takagi, S., Saigusa, T., Ito, K., Bebber, D. P., Fricker, M. D., Yumiki, K., Kobayashi, R., & Nakagaki, T. (2010). Rules for Biologically Inspired Adaptive Network Design. *Science, 327*(5964), 439–442. https://doi.org/10.1126/science.1177894

11) Ben Cranke

12) Fadi El Binni of Al Jazeera Media Network (© CC BY-SA 2.0).
https://www.flickr.com/photos/32834977@N03/5204974979

13) Smithsonian Institution. https://humanorigins.si.edu/

14) José-Manuel Benito Álvarez.
https://www.worldhistory.org/image/17148/oldowan-industry-chopper/

15) Robert G. Bednarik (© CC-BY-SA-4.0).
https://upload.wikimedia.org/wikipedia/commons/thumb/a/ad/Makapansgat_pebble.webp/680px-Makapansgat_pebble.webp.png?20220213115422

16) Sala, N., Arsuaga, J. L., Pantoja-Pérez, A., Pablos, A., Martínez, I. A., Quam, R., Gómez-Olivencia, A., De Castro, J. N., & Carbonell, E. (2015). Lethal Interpersonal Violence in the Middle Pleistocene. *PLOS ONE*, *10*(5), e0126589 (© CC-BY 4.0). https://doi.org/10.1371/journal.pone.0126589

17) Ryan Somma (© CC BY-SA 2.0).

18) NordNordWest. https://upload.wikimedia.org/wikipedia/commons/thumb/2/27/Spreading_homo_sapiens_la.svg/1200px-Spreading_homo_sapiens_la.svg.png

19) Woodman Museum. https://www.facebook.com/WoodmanMuseum/photos/a.780018935383980/2550394531679736/?paipv=0&eav=Afa316eAxc8uRkxQcDBB9ZKcxcehxWjHzMMwiZTmSoQJAO1wzulhhhbtQPmbtiLXgxU&_rdr

Part III

0) Robin-Angelo Photography. Getty Images

1) Childrens Discovery Museum of San Jose. https://www.cdm.org/mammothdiscovery/img/icemaps.gif

2) Cahyo Ramadhani (© CC BY-SA 3.0). https://commons.wikimedia.org/wiki/File:Hands_in_Pettakere_Cave.jpg

3) Stichting Geo Oss. https://www.geo-oss.nl/media/prehistorie.jpg

4) Integral Society https://integralsoc.com/history/hunter-gatherer-society/

5) Ancient Origins https://www.ancient-origins.net/history-important-events/neolithic-revolution-0010298

6) Marie-Lan Nguyen. https://commons.wikimedia.org/wiki/File:Sales_contract_Shuruppak_Louvre_AO3766.jpg

7) Patt O'Neill. http://www.incaglossary.org/appc.html

8) CNG (© CC-BY-SA 3.0). https://en.wikipedia.org/wiki/File:Croeseid_equivalence.jpg

9) Walters Art Museum. https://en.wikipedia.org/wiki/File:Greek_-_Procession_of_Twelve_Gods_and_Goddesses_-_Walters_2340.jpg

10) Jerry Grover (2019). The Swords of Shule. Challex Scientific Publishing. https://www.researchgate.net/figure/Map-of-ancient-Sumer-and-Elam-wwwhyperhistorycom-2016_fig5_334376185

11) Ricardo Liberato (© CC BY-SA 2.0). https://en.wikipedia.org/wiki/File:All_Gizah_Pyramids.jpg

12) Ensinar Historia/Joelza https://www.pinterest.com/pin/796926096536549546/

13) The Minneapolis Institute of Art. https://www.artsmia.org/art-of-asia/history/han-dynasty-map.cfm

14) Cattette (© CC BY 4.0). https://en.wikipedia.org/wiki/File:Achaemenid_Empire_500_BCE.jpg

15) Digital Maps of the Ancient World. https://digitalmapsoftheancientworld.com/digital-maps/the-hellenistic-world/alexander-the-greats-empire/

16) Tristan Hughes, HISTORYHIT. https://www.historyhit.com/how-the-macedonian-phalanx-conquered-the-world/

17) Tataryn (© CC BY-SA 3.0). https://en.wikipedia.org/wiki/File:Roman_Empire_Trajan_117AD.png

18) Geuiwogbil (© CC BY-SA 4.0). https://commons.wikimedia.org/wiki/File:Roman-empire-395AD.svg

19) Arian Zwegers (© CC BY-2.0). https://www.flickr.com/photos/azwegers/15339110573

20) Geoffrey Barraclough. https://en.wikipedia.org/wiki/File:Map_of_expansion_of_Caliphate.svg

21) X Vivid Maps https://vividmaps.com/wp-content/uploads/2015/10/Mongol-Empire.png

22) Highbrow. https://gohighbrow.com/christopher-columbus-1451-1506/

23) The Victoria and Albert Museum. https://artsandculture.google.com/asset/ZgEnAhLJz5AJNA

Part IV

0) National Aeronautics and Space Administration (NASA). https://www.nasa.gov/audience/forstudents/k-4/home/F_Apollo_11.html

1) Bartolomeu Velho (Original work, 1568 CE. Photo taken in 2008 CE). https://en.wikipedia.org/wiki/File:Bartolomeu_Velho_1568.jpg

2) Giuseppe Bertini (1825–1898). https://en.wikipedia.org/wiki/File:Bertini_fresco_of_Galileo_Galilei_and_Doge_of_Venice.jpg

3) Jacob Cornelisz. (Banjaert) van Neck (1564-1638 CE). https://en.wikipedia.org/wiki/File:ThuiskomstVanNeck1599.jpg

4) Vadac. https://en.wikipedia.org/wiki/File:British_Empire_1921.png

5) Nuño García de Toreno: El Planisfero Salviati (1525 CE), University of Seville in Spain. https://grupo.us.es/encrucijada/nuno-garcia-de-toreno-el-planisferio-salviati-1525/

6) Plymouth Chapter of the Society for Effecting the Abolition of the Slave Trade. https://en.wikipedia.org/wiki/File:Slaveshipposter.jpg

7) Roly Williams (© CC BY-SA 3.0). https://commons.wikimedia.org/wiki/File:Oscillating_cylinder.jpg

8) Louis-François, Baron Lejeune (1775–1848 CE). https://en.wikipedia.org/wiki/File:Louis-Fran%C3%A7ois_Baron_Lejeune_001.jpg

9) Adbranch. https://www.adbranch.com/first-steps-of-ford-motor-company-in-business-and-advertising/

10) Émile Alglave & J. Boulard (1884) *The Electric Light: Its History, Production, and Applications*, translated by T. O'Conor Sloan, D. Appleton & Co., New York, p.224, fig.142 on Google Books.

11) Retrieved from Our World Data, by Max Roser, Hannah Ritchie, Esteban Ortiz-Ospina and Lucas Rodés-Guirao (© CC BY 4.0). https://ourworldindata.org/world-population-growth#licence. Modified by Anna de Hek.

12) United States Department of Energy, *Trinity and Beyond: The Atomic Bomb Movie.* https://en.wikipedia.org/wiki/File:Trinity_Detonation_T%26B.jpg

13) National Archives and Records Administration. https://www.archives.gov/research_room/arc/ ARC Identifier: 535916; U.S. Defense Visual Information Center photo HD-SN-99-02993. Retrieved from https://en.wikipedia.org/wiki/File:Rotterdam,_Laurenskerk,_na_bombardement_van_mei_1940.jpg

14) Anna de Hek.

15) National Aeronautics and Space Administration (NASA). https://www.nasa.gov/mission_pages/apollo/apollo-8.html

16) Jpatokal (© CC BY-SA 3.0), (© CC BY-SA 2.5), (© CC BY-SA 2.0), (© CC BY-SA 1.0). https://en.wikipedia.org/wiki/File:World-airline-routemap-2009.png

Part V

0) National Aeronautics and Space Administration (NASA). https://www.nasa.gov/feature/goddard/2022/nasa-s-webb-takes-star-filled-portrait-of-pillars-of-creation

Other Books by
Gerard Alexander Willighagen

Thank You
For Reading

"I'd love to hear from you!"

—Gerard Alexander Willighagen